Operator's Manual For

M1008
M1008A1
M1009
M1010
M1028
M1028A1
M1031

TM 9-2320-289-10
July 1986 edition

edited by
Brian Greul

The M1008 CUCV (Commercial Utility Cargo Vehicle) Series of trucks was produced from 1984 to 1987 by General Motors Defense for the Army, Marines, and Air Force. Approximately 70,000 of these sturdy vehicles were produced. Limited production continued from 1986 to 2001 to accomodate military markets that had the CUCV.

Fonts in this reprint may appear differently due to substitutions for unavailable fonts at the time of printing. A number of black marks were removed from the margins that appeared to be artifacts in the source document, possibly from scanning. Every effort has been made to faithfully reproduce the document while cleaning up the pages to make them usable to you the reader.

Should you have suggestions or feedback on ways to improve this book please send email to Books@OcotilloPress.com

---

Edited 2020 Ocotillo Press
ISBN 978-1-954285-05-7

No rights reserved. This content of this book is in the public domain as it is a work of the US Government. It is reproduced by the publisher as a convenience to enthusiasts and others who may wish to own a quality copy of it. It has been adjusted to accomodate the binding process. The original work was a looseleaf publication in a 3 ring binder.

Printed in the United States of America

Ocotillo Press
Houston, TX 77017
Books@OcotilloPress.com

Disclaimer: The user of this book is responsible for following safe and lawful practices at all times. The publisher assumes no responsibility for the use of the content of this book. The publisher has made an effort to ensure that the text is complete and properly typeset, however omissions, errors, and other issues may exist that the publisher is unaware of.

Removals: The publisher has removed the DA2028-2 Recommended Changes to Technical Publications form from the back of this manual along with the mailng label for such forms that were printed in the original publication.

TM 9-2320-289-10

ARMY TM 9-2320-289-10
AIR FORCE TO 36A12-1A-2081
MARINE CORPS TM 2320-10/1

Supersedes Copy Dated April 1983
See Page i For Details

OPERATOR'S MANUAL FOR

TRUCK, CARGO, TACTICAL, 1-1/4 TON, 4X4, M1008
(2320-01-123-6827)

TRUCK, CARGO, TACTICAL, 1-1/4 TON, 4X4, M1008A1
(2320-01-123-2671)

TRUCK, UTILITY, TACTICAL, 3/4 TON, 4X4, M1009
(2320-01-123-2665)

TRUCK, AMBULANCE, TACTICAL, 1-1/4 TON, 4X4, M1010
(2310-01-123-2666)

TRUCK, SHELTER CARRIER, TACTICAL, 1-1/4 TON, 4X4, M1028
(2320-01-127-5077)

TRUCK, SHELTER CARRIER W/PTO, TACTICAL, 1-1/4 TON, 4X4, M1028A1
(2320-01-158-0820)

TRUCK, CHASSIS, TACTICAL, 1-1/4 TON, 4X4, M1031
(2320-01-133-5368)

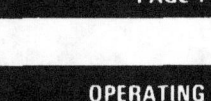
TABLE OF CONTENTS PAGE i

OPERATING PROCEDURES PAGE 2-1

PREVENTIVE MAINTENANCE PAGE 2-5

TROUBLESHOOTING PAGE 3-2

MAINTENANCE PROCEDURES PAGE 3-9

SUBJECT INDEX PAGE INDEX 1

This publication is required for official use or for administrative or operational purposes only. Distribution is limited to U.S. Government Agencies. Other requests for this document must be referred to: Commander, U.S. Army Tank-Automotive Command, ATTN: AMSTA-MBP, Warren, MI 48397-5000.

DEPARTMENTS OF THE ARMY, THE AIR FORCE, AND HEADQUARTERS, MARINE CORPS

JULY 1986

TM 9-2320-289-10
TO 36A12-1A-2081 TM
2320-1011

C 5

CHANGE

No. 5

DEPARTMENT OF THE ARMY, THE AIR FORCE, AND HEADQUARTERS, MARINE CORPS Washington, D.C., 1 May 1992

OPERATOR'S MANUAL
FOR

TRUCK, CARGO, TACTICAL, 1-1/4 TON, 4X4, M1008
(2320-01-123-6827)

TRUCK, CARGO, TACTICAL, 1-1/4 TON, 4X4, M1008A1
(2320-01-123-2671)

TRUCK, UTILITY, TACTICAL, 3/4 TON, 4X4, M1009
(2320-01-123-2665)

TRUCK, AMBULANCE, TACTICAL, 1-1/4 TON, 4X4, M1010
(2310-01-123-2666)

TRUCK, SHELTER CARRIER, TACTICAL, 1-1/4 TON, 4X4, M1028
(2320-01-127-5077)

TRUCK, SHELTER CARRIER W/PTO, TACTICAL, 1-1/4 TON, 4X4, M1028A1
(2320-01-158-0820)

TRUCK, SHELTER CARRIER W/PTO, TACTICAL 1-1/4 TON, 4X4, M1028A2
(2320-01-295-0822)

TRUCK, SHELTER CARRIER, TACTICAL, 1-1/4 TON, 4X4, M1028A3
(2320-01-325-1937)

TRUCK, CHASSIS, TACTICAL, 1-1/4 TON, 4X4, M1031
(2320-01-133-5368)

TM 9-2320-289-10, dated 4 July 1986, is changed as follows:

1. The manual title is changed to read as shown above.

2. Remove old pages and insert new pages.

3. New or changed material is indicated by a vertical bar in the margin of the page and by a vertical bar adjacent to the TA number.

Remove Pages Insert Pages

i through iiil(iv blank) i through iiil(iv blank) 1-1 through 1-4 1-1 through 1-4 1-7 and 1-8
1-7 and 1-8
1-12.1/(1-12.2blank) 1-12.1 and 1-12.3/(1-12.4blank) 1-19 and 1-20 1-19 and 1-20
2-19 and 2-20 2-19 and 2-20

Distribution authorized to U.S. Government agencies for administrative end operational purpose only. Other requests for this document will be referred to: Commander, U.S. Army Tank-Automotive Command, ATTN: AMSTA-MB, Warren, MI 48397-5000 .

Destromy by any method that will prevent disclosure of contents or reconstruction of the docu-ment .

1

Remove Pages (Con't)	Insert Pages (Con't)
2-23 and 2-24	2-23 and 2-24
2-29 and 2-30	2-29 and 2-30
2-53 and 2-54	2-53 and 2-54
2-57 and 2-58	2-57 and 2-58
2-63 and 2-64	2-63 and 2-64
2-81 and 2-82	2-81 and 2-82
2-89 and 2-90	2-89 and 2-90
3-5 and 3-6	3-5 and 3-6
3-11 and 3-11.0/(3-11.1 blank) 3-13 through 3-15/ (3-16 blank)	3-11 and 3-11.0 /(3-11.1 blank) 3-13 through 3-15/(3-16 blank)
B-31 (B-4 blank)	B-3/(B-4 blank)
C-1 and C-2	C-1 and C-2

4. File this change sheet in front of the publication for reference purposes. By Order of the Secretary of

GORDON R. SULLIVAN
General, United States Army
Chief of Staff

Official:

MILTON H. HAMILTON
Administrative Assistant to the
Secretary of the Army
02692

By Choler of the Secretary of the Air Force:

MERRILL A. McPEAK
General, United States Air Force
Chief of Staff

CHARLES C. MCDONALD
General, United States Air Force
Commander, Air Force Logistics Command

By Order of the Marine Corps:

H.E. REESE
Deputy for Support
Marine Corps Research, Development and
Acquisition Command

Distribution:

To be distributed in accordance with DA Form 12-38-E (Block 0369) Operator mainte-nance requirements for TM9-2320-289-10.

TM 9-2320-289-10
TO 36A12-1A-2081
TM 2320-10/1
C4

CHANGE

NO. 4

DEPARTMENT OF THE ARMY, THE AIR FORCE,
AND HEADQUARTERS, MARINE CORPS

Washington D.C., *31 January 1992*

OPERATOR'S MANUAL
FOR

TRUCK, CARGO, TACTICAL, 1-1/4 TON, 4X4, M1008
(2320-01-123-6827)

TRUCK, CARGO, TACTICAL, 1-1/4 TON, 4X4, M1008A1
(2320-01-123-2671)

TRUCK, UTILITY, TACTICAL, 3/4 TON, 4X4, M1009
(2320-01-123-2665)

TRUCK, AMBULANCE, TACTICAL, 1-1/4 TON, 4X4, M1010
(2310-01-123-2666)

TRUCK, SHELTER CARRIER, TACTICAL, 1-1/4 TON, 4X4, M1028
(2320-01-127-5077)

TRUCK, SHELTER CARRIER W/PTO, TACTICAL, 1-1/4 TON, 4X4, M1028A1
(2320-01-158-0820)

TRUCK, SHELTER CARRIER W/PTO, TACTICAL 1-1/4 TON, 4X4, M1028A2
(2320-01-295-0822)

TRUCK, CHASSIS, TACTICAL, 1-1/4 TON, 4X4, M1031
(2320-01-133-5368)

TM 9-2320-289-10, dated 4 July 1986, is changed as follows:

1. Remove old pages and insert new pages as indicated below.
2. New or changed material is indicated by a vertical bar in the margin of the page.
3. The Preventive Maintenance Checks and Services have been completely replaced; no change bars or pointing hands will appear on pages 2-8 through 2-49

Remove Pages	Insert Pages
2-7 through 2-52	2-7 through 2-49/(2-50 blank)

4. File this change sheet in front of the publication for references purposes.

This publication is required for official use or for administrative or operational purposes only. Distribution is limited to U.S. Government Agencies. Other requests for this document must be referred to: Commander, U.S. Army Tank–Automotive Command, TTN: AMSTA–MBP, Warren, MI 48397–5000.

Destroy by any method that will prevent disclosure of contents or reconstruction of the document.

By Order of the Secretary of the Army:

                        GORDON R. SULLIVAN
                      General, United States Army
                              Chief of Staff

Official:

MILTON H. HAMILTON
Administrative Assistant to the
Secretary of the Army

00563

By Order of the Secretary of the Air Force:

                        MERRILL A. McPEAK
                            General USAF
                              Chief of Staff

CHARLES C. McDONALD
General, United States Air Force
Commander, Air Force Logistics Command

By Order of the Marine Corps:

                            H. E. REESE
                           Deputy for Support
           Marine Corps Research, Development and
                       Acquisition Command

Distribution:
   To be distributed in accordance with DA Form 12-38-E, Block 0369, Operator maintenance requirements for TM 9-2320-289-10.

TM 9-2320-289-10
TO 36A12-1A-2081
TM 2320-1011
C 3

CHANGE

NO. 3

DEPARTMENT OF THE ARMY, THE AIR FORCE,
AND HEADQUARTERS, MARINE CORPS
Washington D.C., *28 December 1990*

OPERATOR'S MANUAL
FOR

TRUCK, CARGO, TACTICAL, 1-1/4 TON, 4X4, M1008
(2320-01-123-6827)

TRUCK, CARGO, TACTICAL, 1-1/4 TON, 4X4, M1008A1
(2320-01-123-2671)

TRUCK, UTILITY, TACTICAL, 3/4 TON, 4X4, M1009
(2320-01-123-2665)

TRUCK, AMBULANCE, TACTICAL, 1-1/4 TON, 4X4, M1010
(2310-01-123-2666)

TRUCK, SHELTER CARRIER, TACTICAL, 1-1/4 TON, 4X4, M1028
(2320-01-127-5077)

TRUCK, SHELTER CARRIER W/PTO, TACTICAL, 1-1/4 TON, 4X4, M1028A1
(2320-01-158-0820)

TRUCK, SHELTER CARRIER W/PTO, TACTICAL 1-1/4 TON, 4X4, M1028A2
(2320-01-295-0822)

TRUCK, CHASSIS, TACTICAL, 1-1/4 TON, 4X4, M1031
(2320-01-133-5368)

TM 9-2320-289-10, dated 4 July 1986, is changed as follows:

1. The manual title is changed to read as shown above.

2. Remove old pages and insert new pages.

3. New or changed material is indicated by a vertical bar in the margin of the page and by a vertical bar adjacent to the TA number.

Remove Pages	Insert Pages
i through iii/(iv blank)	i through iii/(iv blank)
1-1 through 1-4	1-1 through 1-4
1-7 and 1-8	1-7 and 1-8
None	1-12.1/(1-12.2 blank)
1-15 and 1-16	1-15 and 1-16
1-19 and 1-20	1-19 and 1-20
2-19 and 2-20	2-19 and 2-20

Remove Pages (Con't)	Insert Pages (Con't)
2-57 and 2-58	2-57 and 2-58
2-81 and 2-82	2-81 and 2-82
2-89 and 2-90	2-89 and 2-90
3-11 and 3-12	3-11 through 3-12
A-1/(A-2 blank)	A-1/(A-2 blank)
B-3/(B-4 blank)	B-3/(B-4 blank)
C-1 and C-2	C-1 and C-2
E-3 and E-4	E-3 and E-4
Index 3 and Index 4	Index 3 and Index 4

4. File this change sheet in front of the publication for reference purposes.

By Order of the Secretary of the Army:

CARL E. VUONO
General, United States Army
Chief of Staff

Official:

THOMAS F. SIKORA
Brigadier General, United States Army
The Adjutant General

By Order of the Secretary of the Air Force:

LARRY D. WELCH
General, United States Air Force
Chief of Staff

CHARLES C. McDONALD
General, United States Air Force
Commander, Air Force Logistics Command

By Order of the Marine Corps:

H. E. REESE
Deputy for Support
Marine Corps Research, Development and
Acquisition Command

Distribution:

To be distributed in accordance with DA Form 12-38-E, Block 0369, Operator maintenance requirements for TM 9-2320-289-10.

CHANGE

NO. 2

TM 9-2320-289-10
TO 36A12-1A-2081
TM 2320-10/1
C 2
HEADQUARTERS
DEPARTMENT OF THE ARMY
Washington, D.C., 16 October 1989

OPERATOR'S MANUAL
FOR

TRUCK, CARGO, TACTICAL, 1-1/4 TON, 4X4, M1008
(2320-01-123-6827)

TRUCK, CARGO, TACTICAL, 1-1/4 Ton, 4X4, M1008A1
(2320-01-123-2671)

TRUCK, UTILITY, TACTICAL, 3/4 TON, 4X4, M1009
(2320-01-123-2665)

TRUCK, AMBULANCE, TACTICAL, 1-1/4 TON, 4X4, M1010
(2310-01-123-2666)

TRUCK, SHELTER CARRIER, TACTICAL, 1-1/4 Ton, 4X4, M1028
(2320-01-127-5077)

TRUCK, SHELTER CARRIER W/PTO, TACTICAL, 1-1/4 TON, 4X4, M1028A1
(2320-01-158-0820)

TRUCK, CHASSIS, TACTICAL, 1-1/4 TON, 4X4, M1031
(2320-01-133-5368)

TM 9-2320-289-10, 4 July 86, is changed as follows:

1. Remove old pages and insert new pages as indicated below.

2. New or changed material is indicated by a vertical bar in the margin of the page.

REMOVE PAGES	INSERT PAGES
1-19 and 1-20	1-19 and 1-20

3. File this change sheet in front of the publication for reference purposes.

By Order of the Secretary of the Army:

CARL E. VUONO
General, United States Army
Chief of Staff

Official:

WILLIAM J. MEEHAN II
Brigadier General, United States Army
The Adjutant General

By Order of the Secretary of the Air Force:

LARRY D. WELCH
General, United States Air Force
Commander, Air Force Logistics Command

ALFRED G. HANSEN
General, United States Air Force
Commander, Air Force Logistics Command

By Order of the Marine Corps:

H.E. REESE
Deputy for Support
Marine Corps Research, Development and
Acquisition Command

Distribution:

To be distributed in accordance with DA Form 12-38, Operator maintenance requirements for Truck, Commercial Utility Vehicle, Cargo, Tactical, 4x4, M1008, M1008A1, M1009, M1010, M1028, M1028A1, M1031.

CHANGE	TM 9-2320-289-10
	TO 36A12-1A-2081
NO. 1	TM 2320-10/1
	C 1
	DEPARTMENT OF THE ARMY,
	AIRFORCE AND HEADQUARTERS
	MARINE CORPS
	Washington D.C., *27 May 1987*

OPERATOR'S MANUAL
FOR

TRUCK, CARGO, TACTICAL, 1-1/4 TON, 4X4, M1008
(2320-01-123-6827)

TRUCK, CARGO, TACTICAL, 1-1/4 Ton, 4X4, M1008A1
(2320-01-123-2671)

TRUCK, UTILITY, TACTICAL, 3/4 TON, 4X4, M1009
(2320-01-123-2665)

TRUCK, AMBULANCE, TACTICAL, 1-1/4 TON, 4X4, M1010
(2310-01-123-2666)

TRUCK, SHELTER CARRIER, TACTICAL, 1-1/4 Ton, 4X4, M1028
(2320-01-127-5077)

TRUCK, SHELTER CARRIER W/PTO, TACTICAL, 1-1/4 TON, 4X4, M1028A1
(2320-01-158-0820)

TRUCK, CHASSIS, TACTICAL, 1-1/4 TON, 4X4, M1031
(2320-01-133-5368)

TM 9-2320-289-10, 4 July 86, is changed as follows:

1. Remove old pages and insert new pages as indicated below.

2. New or changed material is indicated by a vertical bar in the margin of the page.

REMOVE PAGES	INSERT PAGES
Warning a. and Warning b.	Warning a. and Warning b.
1-3 and 1-4	1-3 and 1-4
	New page 2.56.1/(2.56-2 blank)

3. File this change sheet in front of the publication for reference purposes.

TM 9-2320-289-10

By Order of the Secretaries of the Army, the Navy, and the Air Force:

JOHN A. WICKHAM, JR.
General, United States Army
Chief of Staff

Official:

R.L. DILWORTH
Brigadier General, United States Army
The Adjutant General

Official:

NORMAND G. LEZY, Colonel, USAF
Director of Administration

LARRY D. WELCH, General, USAF
Chief of Staff

Official:

J.J. WENT
Lieutenant General, USMC
Deputy Chief of Staff for Installations and Logistics

Distribution:

To be distributed in accordance with DA Form 12-38, Truck, Commercial Utility Vehicle, Cargo, Tactical, 4x4, M1008, M1008A1, M1009, M1010, M1028, M1031.

TM 9-2320-289-10

WARNING

EXHAUST GASES CAN KILL!

1. DO NOT operate truck engine in enclosed area.
2. DO NOT idle truck engine with cab windows closed.
3. DO NOT drive truck with inspection plates or cover plates removed
4. BE ALERT at all times for exhaust odors.
5. BE ALERT for exhaust poisoning symptoms. They are:

    Headache

    Dizziness

    Sleepiness

    Loss of Muscular Control

6. If you see a person with exhaust poisoning symptoms:

    Remove person from area.

    Expose to open air.

    Keep person warm.

    Do not permit person to move.

    Administer artificial respiration, if necessary. "

    Immediately notify medical personnel.

WARNING

PARKING

Whem leaving the truck:

   DO NOT use transmission as a substitute for parking brake. ALWAYS engage parking brake when parking truck.

   If transfer case control lever is in "N" (Neutral), transfer case is disengaged and shifting transmission gearshift lever to "P"(Park) WILL NOT stop the truck from moving.

   Turn ignition key to   LOCK Remove key.

Failure to follow this warning can result in injury to personnel or equipment damage.

*For First Aid . refer to FM 21-11.

Warning a

TM 9-2320-289-10

WARNING

EXHAUST PIPE AND MUFFLER

- DO NOT park or idle truck over combustible materials such as grass or leaves, if tactical situation permits. They could ignite from the heat of the exhaust system and start a fire. Failure to follow this warning can result in injury to personnel or equipment damage.

- DO NOT touch hot exhaust pipes or muffler with bare hands. Severe injury can result.

- If the tailgate and/or tailgate window is required to be opened while moving or towing, extreme caution is required. Exhaust fumes may enter resulting in injury or death to personnel.

WARNING

DIESEL FUEL HANDLING

DO NOT SMOKE OR PERMIT ANY OPEN FLAME IN AREA OF TRUCK WHILE. YOU ARE SERVICING DIESEL FUEL SYSTEM. Be sure hose nozzle is grounded against filler tube during refueling to prevent static electricity. Failure to follow this warning can result in injury to personnel or equipment damage.

WARNING

BATTERIES

Remove all jewelry such as rings, dog tags, bracelets, etc. If jewelry contacts battery terminal, a direct short

WARNING

CLEANING AGENTS

DO NOT SMOKE when using cleaning solvent. NEVER USE IT NEAR AN OPEN FLAME. Be sure there is a fire extinguisher nearby and use cleaning solvent only in well-ventilated places. Flash point of solvent is

138°F (60°C).
USE CAUTION when using cleaning solvents. Cleaning solvents evaporate quickly and can irritate exposed skin if solvents contact skin. In cold weather, contact of exposed skin with cleaning solvents can cause

WARNING

GAS-PARTICULATE FILTER UNIT

Under extreme cold conditions, danger of frostbite exists. Put on protective mask, but DO NOT connect air duct to mask until heater has been on for 15 minutes.

*TM 9-2320-289-10

TECHNICAL MANUAL
TM 9-2320-289-10

DEPARTMENTS OF THE ARMY, THE AIR FORCE, AND
HEADQUARTERS, MARINE CORPS Washington,
D.C. *4 July 1986*

OPERATOR'S MANUAL
FOR

TRUCK, CARGO, TACTICAL, 1-1/4 TON, 4X4, Ml 008
(2320-01-123-6827)

TRUCK, CARGO, TACTICAL, 1-1/4 TON, 4X4, M1008A1
(2320-01-123-2671)

TRUCK, UTILITY, TACTICAL, 3/4 TON, 4X4, M1009
(2320-01-123-2665)

TRUCK AMBULANCE, TACTICAL, 1-1/4 TON, 4X4, M1010
(2310-01-123-2666)

TRUCK, SHELTER CARRIER, TACTICAL, 1-1/4 TON, 4X4, M1028
(2320-01-127-5077)

TRUCK SHELTER CARRIER W/PTO, TACTICAL, 1-1/4 TON, 4X4, M1028A1
(2320-01-158-0820)

TRUCK, SHELTER CARRIER W/PTO, TACTICAL, 1-1/4 TON, 4X4, M1028A2
(2320-01-295-0822)

TRUCK, SHELTER CARRIER, TACTICAL, 1-1/4 TON, 4X4, M1028A3
(2320-01-325-1937)

TRUCK, CHASSIS, TACTICAL, 1-1/4 TON, 4X4, M1031
(2320-01-133-5368)

REPORTING ERRORS ANO RECOMMENDING IMPROVEMENTS

(Army) You can improve this manual. If you find any mistakes or if you know of a way to improve the procedures, please let us know. Mail your letter, DA Form 2028 (Recommended Changes to Publications and Blank Forms), or DA Form 2028-2, located in the back of this manual, direct to Commander, U.S. Army Tank-Automotive Command, ATTN: AMSTA-MB, Warren, MI 48397-5000. (Marine Corps) Submit NAVMC 10772 to the Commanding General, Marine Corps Logistic Base (Code 850), Albany, GA 31704. A reply will be furnished to you.

TABLE OF CONTENTS

	Page
How To Use This Manual	iii
CHAPTER 1 INTRODUCTION	1-1
Section I General Information	1-1
Section II Equipment Description	1-3
Section III Technical Principles of Operation	1-13

*This publication supersedes TM 9-2320-289-10 dated April 1983.

TM 9-2320-289-10

TABLE OF CONTENTS - Continued

Page

CHAPTER 2 OPERATING PROCEDURES 2-1

    Section I Description and Use of Controls and Indicators 2-1

    Section II Preventive Maintenance Checks and Services (PMCS) 2-5

    Section III Operate Under Usual Conditions 2-53

    Section IV Operate Ambulance Peculiar Components 2-65

    Section V Operate Auxiliary Equipment ...2-80

    Section VI Operate Under Unusual Conditions 2-86

CHAPTER 3 MAINTENANCE INSTRUCTIONS 3-1

    Section I Lubrication 3-1

    Section II Troubleshooting 3-2

    Section III Maintanance Procedures........................... .. 3-9

    Section IV Maintenance Under Unusual Conditions 3-14

APPENDIX A REFERENCES A-1

APPENDIX B COMPONENTS OF END ITEM AND BASIC ISSUE ITEMS LISTS B-1

APPENDIX C ADDITIONAL AUTHORIZATION LIST C-1

APPENDIX D EXPENOABLE/OURABLESUPPLIESANO MATERIALS LIST D-1

APPENDIX E STOWAGE ANO SIGN GUIDE FOR COMPONENTS OF END ITEM, BASIC
    ISSUE ITEMS, AND APPLICABLE ADDITIONAL AUTHORIZATION ITEMS E-1

Index 1

TM 9-2320-289-10

HOW TO USE THIS MANUAL

This manual Is designed to help operate and maintain the M1008, M1008A1, M1009, M1010, M1028, M1028A1, M1028A2, M1028A3, and M1031 trucks. Listed below are special features which trove been Included to make it easier to locate and to use the Information You need.

- A table of contents is provided, giving e quick reference to chapters snd sections that will be used often.

- Warnings, subject headings, procedure steps and certain other kinds of information are highlighted in bold print as a visual aid.

All capital letters are used to emphasize statements of particular importance. FOLLOW THESE GUIDELINES WHEN YOU USE THIS MANUAL

Read all warnings, cautions, and notes.

- Within a chapter or section, headings are used to help group the material and to assist you in quickly finding tasks.

- The operator should read through this manual and become familiar with the contents before attempting to operate the truck.

Technical instructions will include metric units in addition to nonmetric units.

Change 5 iii/(iv blank)

TM 9-2320-289-10

CHAPTER 1

INTRODUCTION

Section I. GENERAL INFORMATION

1-1. SCOPE

This manual is for your use In operating, troubleshooting, and maintaining the Ml 008, M1008A1, M1010, M1028, M1028A1, M1028A2, M1028A3, and M1031 1-1/4 ton trucks and the M1009 3/4 ton truck. It also provides instructions for trucks equipped with special purpose kits.

1-2. MAINTENANCE FORMS AND RECORDS

Department of the Army forms and procedures used for equipment maintenance will be those prescribed by DA Pam 738-750, the Army Maintenance Management System (TAMMS).

1-3. REPORTING EQUIPMENT IMPROVEMENT RECOMMENDATIONS (EIRs)

If your Commercial Utility Cargo Vehicle (CUCV) Series truck needs improvement, let us know. Send us an EIR. You, the user, are the only one who can tell us what you don't like about your equipment. Let us know why you don't like the design or performance. Put it on an SF 368 (Quality Deficiency Report). Mail it to us at Commander, U.S. Army Tank-Automotive Command, ATTN: AMSTA-QRT, Warren, MI 48397-5000. We'll send you a reply.

1-4. WARMNING INFORMATION

The M1008, M1008A1, M1009, M1010, M1028, M1028A1, M1028A2, M1028A3, and M1031   truck are under warranty by the Chevrolet Motor Division of the General Motors Corporation In accordance with TB 9-2300-295-15/24.

1-5. ABBREVIATIONS

Listed below are explanations of abbreviations used on the truck data service plates, and throughout this manual. "

A.	After (PMCS)
AAL	Additional Authorization List
B.	Before (PMCS)
Bli	Basic Issue Items
C.	Centigrade or Celsius operator/crew (maintenance level) centimeter
C.	
cm	Components of End Item
COED	Common Table of Allowances
CTA	Drive (on transmission gearshift lever)
D.	
D.	During (PMCS)
DA	Department of the Army
EIR	Equipment Improvement Recommendation
F.	Fahrenheit
GAWR	Gross Axle Weight Rating
GVWR	Gross Vehicle Weight Rating
JTA	Joint Table of Allowance kilogram
kg.	km kilometer kilopascal
	kPa
	kph kilometers per hour liter
l.	
m.	meter
M	Monthly (PMCS)
mm	millimeter
N.	Neutral (on transmission gearshift lever)

Change 5   1-1

TM 9-2320-289-10

NSN	National Stock Number
P	Park (on transmission gearshift lever)
PMCS	Preventive Maintenance Checks and Services
R	Reverse (on transmission gearshift lever)
T/C	Transfer Case
TDA	Table of Distribution and Allowances
TOE/MTOE	Table of Organizational Equipment/Modified Table of Organizational Equipment"
USATACOM	U.S. Army Tank-Automotive Command
W	Weekly (PMCS)

TM 9-2320-289-10

Section II. EQUIPMENT DESCRIPTION

1-6. EQUIPMENT CHARACTERIS,CAS, CAPABILITIES, ANO FEATURES

a. The trucks in this series are commercial trucks suitable for use on all types of roads and highways, in all types of weather. The trucks are further designed for infrequent off-road operations.

b. The trucks are capable of occasional hardbottom fording to a depth of 20 inches(51 cm) at 5 mph (8 kph) for not more than approximately 3 minutas without stalling the engine, causing permanent damage to components, or requiring immediate maintenance.

c. Features of the truck series include:

(1) automatic transmission with three forward and one reverse spaeds 379 cubic
(2) inch (6.2 liter) diesel V-8 engine
(3) manually activated transfer case for four-wheel drive operations
(4) hydraulically activated, power assisted front disc and rear drum service brakes NATO
(5) slave cable receptacle for slave starting 1 2/28 volt electrical system
(6) winterization kit available for each model
(7) d. Cargo Truck, The M1008, M1008A1, M1028, M1028A1, M1028A2, and

M1028A3 1-1/4 ton, 4X4, cargo trucks are. fight commercial trucks designed to provide standard tactical mo-bility and to carry cargo or passengers, All models have a tow pintle at the rear bumper to permit towing of trailer or aircraft. The M1008 can be equipped with a troop aeat kit for eight personnel. The M1008A1 Includes a 100 amp/28 v communlcations kit. The M1028 is equipped to accept a 100 amp/28 v electrical system and S250 communications shelter and a communication kit. The M1028A1 Is equipped wfth a New Process 205 Transfer Case which allows for the addition of a power take-off (PTO) unit. The M1028A2 is an M1028A1 that hae been converted to rear dual wheel conflg-uration. The M1028A3 is an M1028 whkh has been converted to rear dual wheel shelter carrier configuratlon and is equipped with a New Process 208 Transfer Case.

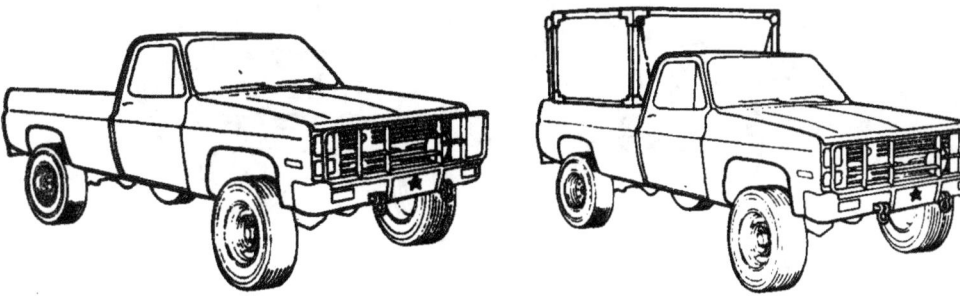

M1008   **M1028**

Change 5 1-3

TM 9-2320-289-10

e. Utility Truck. The M1009 is a 3/4 ton truck that has an enclosed body and can be used for command and control purposes.

M1009

f. Ambulance. The M1010 Ambulance is designed to carry four litter or eight ambulatory patients with both upper litters in place; or two litter and four ambulatory patients with one upper litter in place, It is equipped with a 200 amp/28 v electrical system, which supports a gas-particulate filter system, a patient compartment air conditioner, and a fuel fired patient compartment heater.

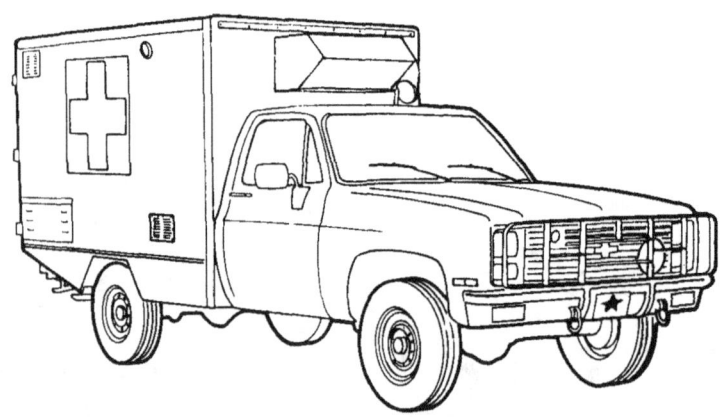

M1010

TA466565

1-4

g. Chassis Truck. The M1031 is a chassis and cab combination, which was designed for the mounting of special bodies that might be required.

M1031

TM 9-2320-289-10

1-7. LOCATION AND DESCRIPTION OF MAJOR COMPONENTS
　　a. External Components.

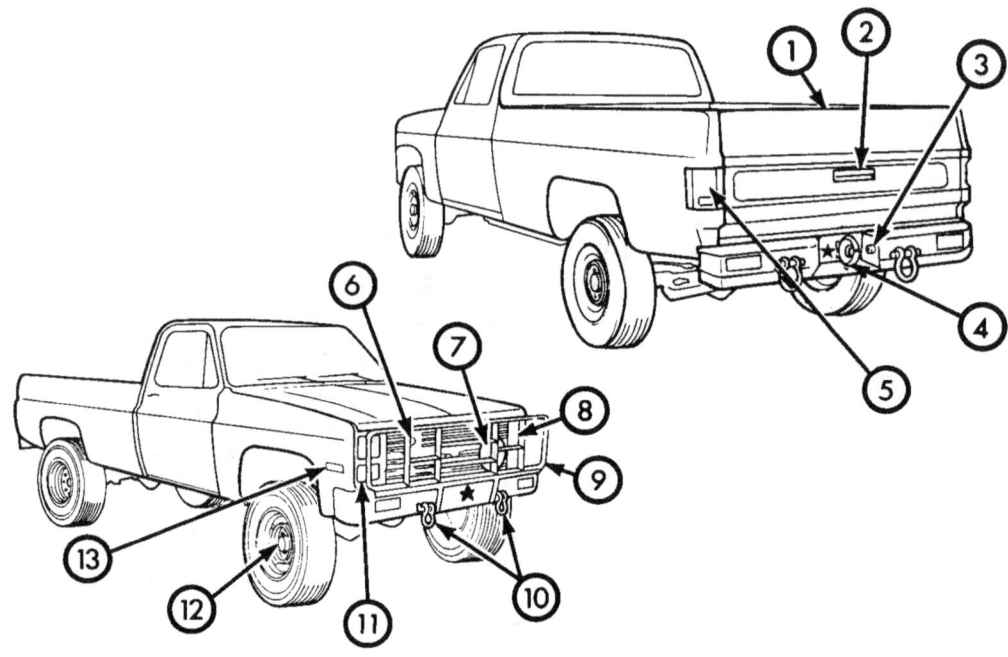

M1008

1. Tailgate.

2. Tailgate Latch Releasa. Releases tailgate latch, allowing tailgate to be lowered.

3. Trailer Electrical Coupling. Trailer lights cable to be hooked here during trailer towing.

　Tow Pintle. Used to tow trailers or aircraft.
4.
　Stoplights/Taillights.
5.
　Slave Recaptacle. Used to start truck with another vehicle having a 28 volt starting system.

6. Weight Classification Marker.

7. Blackout Drive Light. Used during hours of limited visibility when service lights cannot be used.

8. Brush Guard. Protects radiator from damage,

1. Tow Hooks. Used to tow truck short distances.
9.
2. Service Lights.

3. Locking Hubs. Used to prepare truck for four-wheel drive operation,

4. Sida Markers.

TA466567

1-6

TM 9-2320-289-10

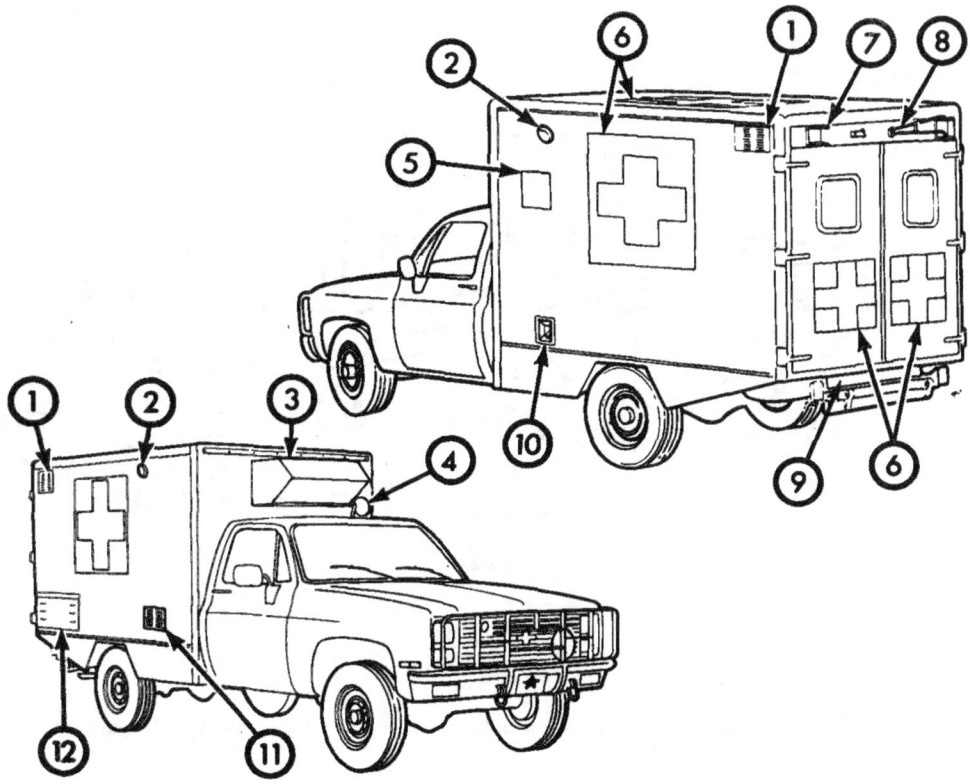

M1010

1. Air Exhaust Vent. Allows air to leave patient compartment.

2. Floodlights. One on each side of ambulance body. Can be adjusted to meet mission requirements.

3. Air Conditioner. Provides climate control of patient compartment during hot weather.

4. Spotlight. Spotlight can be directed as desired from inside cab by twisting and rotating handle.

5. Medical Corps Symbol.

6. Red Cross Symbol. Removable; one on each side and on top, and two on rear doors.

7. Patient Assist Boom Bracket. To secure boom to other side of doors.

8. Patient Assist Boom. Used to assist in lifting patients into upper litter berths.

9. Access Steps. Foldout patient compartment steps allow personnel to easily enter and exit patient compartment.

10. Fuel Fillar Cap.

11. Heater Air Inlet. Allows air to enter for personnel heater.

Heater Access Door. Allows access to personnel heater for maintenance.

TM 9-2320-289-10

M100 9

1. Tailgate Window.

2. Tailgate Latch Release. Located on inside, just below window. Releases tailgate latch allowing tailgate to be lowered.

3. Tailgate Window Crank. Raises and lowers window in utility tailgate.

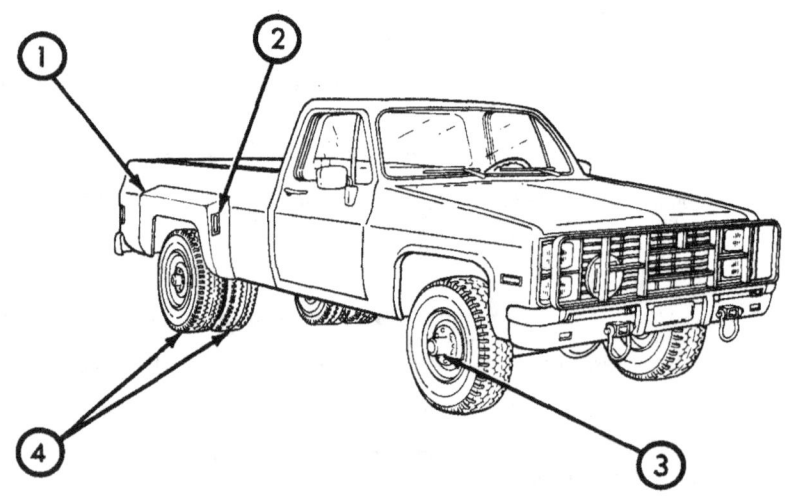

M1028A2 AND M1028A3

1. Flared Rear Fender.

2. Rear Fender Side Markar.

3. Offset Front Wheel Hub.

4. Dual Rear Wheels.

b. Internal Components.

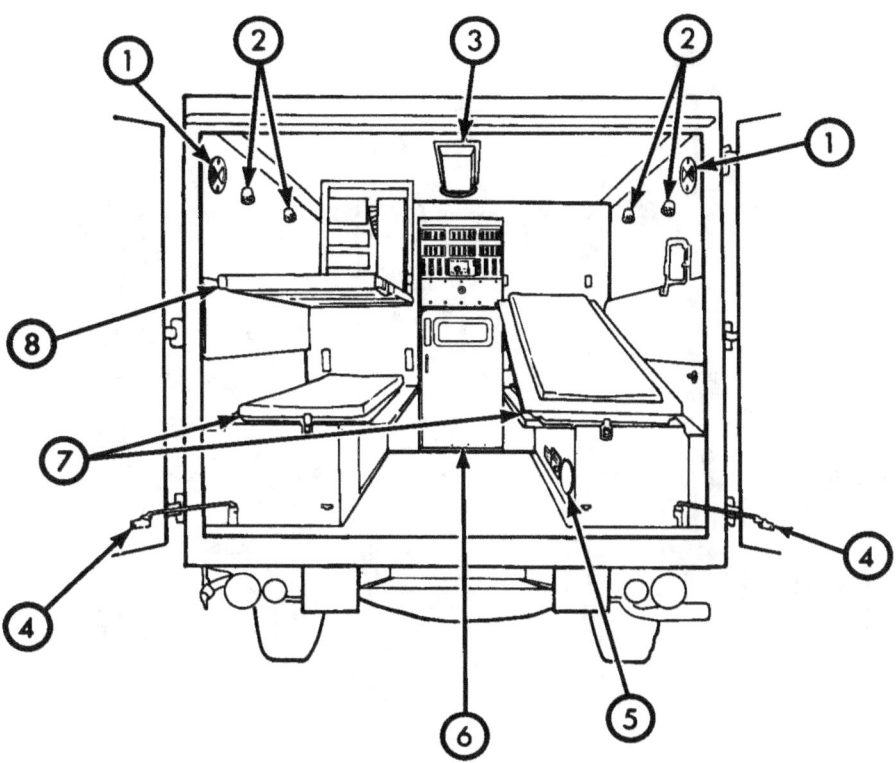

M1010 Ambulance Patient Compartment

1. Air Exheust Vents. Allow air to leave patient compartment.
2. Focus Lights. Provide direct light for specific applications; for medical use.
3. Domelight. Provides overhead light for working within patient compartment.
4. Rear Door Hold Open. Holds patient compartment doors in open position.
5. Heater Outlet. Allows warm air from heater to enter patient compartment.
6. Patient Compartment Front Door. Allows access to
7. Lower Litter Berths.
8. Upper Litter Berths.

TM 9-2320-289-10

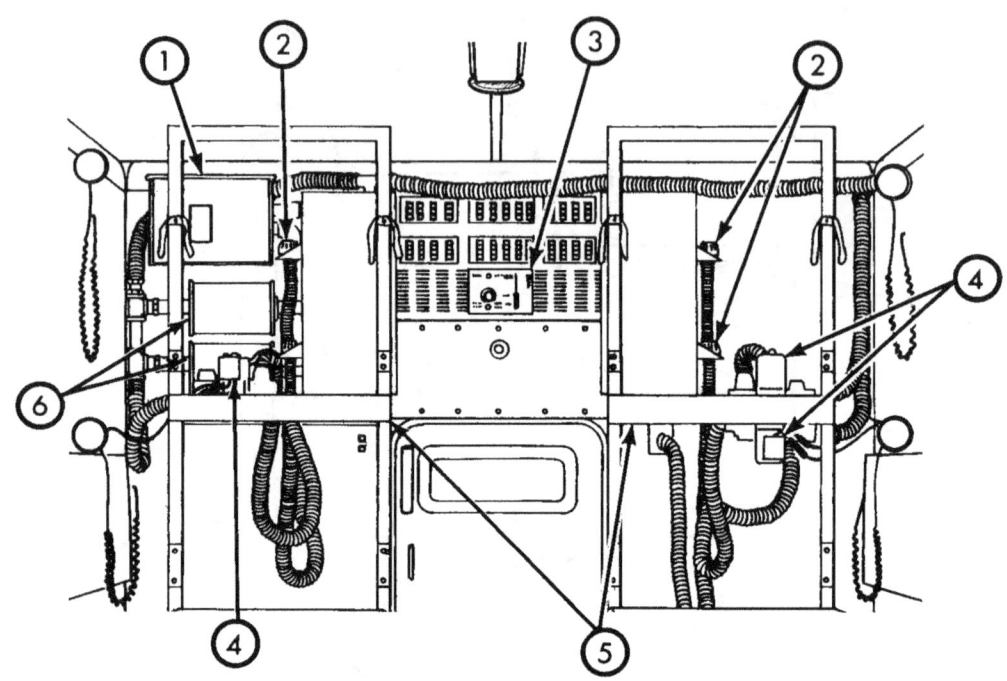

M1010 Ambulance Patient Compartment

1. Gas-Particulate Filter Unit (G PFU). Pumps and filters sir from patient compartment.

2. Gas-Particulate Filter Air Outlet. Connects to protective mask and delivers heated and filtered air to wearer of mask,

3. Air Conditional Controls. Regulate air flow through air conditioner unit and control temperature.

4. Gas-Particulate Filter Unit Heater. Heats air from gas-particulate filter unit during cold weather. There are five heaters in truck; three in patient compartment and two in cab.

Upper Litter Berth Support. Supports upper litter berth,

5. Gs-Particulate Filter Canisters. Provide a second filter for air passing through filter unit.

6.

1-10

1-8. DIFFERENCES BETWEEN MODELS

	M1008	R1008A1	M1009	M1010	M1028	M1028A1	M1031
a. 6.2 Liter Diesel Engine	x	x	x	x	x	x	x
b. Automatic Transmission	x	x	x	x	x	x	x
c. Model 208 Transfer Case (T/C)	x	x	x	x	x		x
d. Model 205 Transfer Case (T/C)						x	
e. 100 amp/28 volt Electrical System			x				
f. 200 amp/28 volt Electrical System	x	x		x	x	x	x
g. Troop Seat Kit					x	x	
h. Communications Kit				x			
i. Slave Receptacle	x	x					
j. Tow Pintle		x	x		x	x	
k. Cargo Tie-downs	x	x	x	x	x	x	x
l. Shelter Tie-downs	x	x	x				x
m. Air Conditioner	x	x					
n. Gas-Particulate Filter System					x	x	
o. Locking Differential				x			
p. No-Spin Differential				x			
q. Cargo Cover Kit	xx	xx		x	x	x	x
r. Rear Passenger Seating			xx				
s. Spotlight							
t. Floodlights				xx			

1-9. TABULATED DATA

Basic information you will need to know about the trucks is contained in Table 1-1.

Table 1-1. Tabulated Data

DATA	MODEL				
	M 1008 M1008A1	M1009	M1010	M1028 MA1028A1	M1031
Make	Chevrolet	Chevrolet	Chevrolet	Chevrolet	Chevrolet
Weight:					
Curb	5900 lb (2679 kg)	5200 lb (2361 kg)	7370 lb (3346 kg)	5800 lb (2633 kg)	5250 lb (2384 kg)
Payload/Passengers	2900 lb (1317 kg)	1200 lb (545 kg)	2080 lb (944 kg)	3600 lb (1634 kg)	3950 lb (1793 kg)
GVWR	8800 lb (3995 kg)	6400 lb (2906 kg)	9450 lb (4290 kg)	9400 lb (4268 kg)	9200 lb max (4177 kg)
GAWR (front)	4500 lb (2043 kg)	3600 lb (1634 kg)	4500 lb (2043 kg)	4500 lb (2043 kg)	4500 lb (2043 kg)
GAWR (rear)	7000 lb (3178 kg)	3750 lb (1703 kg)	7000 lb (3178 kg)	7500 lb (3405 kg)	7500 lb [3405 kg]
Wheelbase	131.5 in (334 cm)	106.5 in (271 cm)	131.5 in (334 cm)	131.5 in (334 cm)	131.5 in (334 cm)
Track (front)	67.8 in (172 cm)	68 in (173 cm)	67.8 in (172 cm)	67.8 in (172 cm)	67.8 in [172 cm]
Track (rear)	65.8 in (167 cm)	65 in (165 cm)	65.8 in (167 cm)	65.8 in (167 cm)	65.8 in [167 cm]

Table 1-1. Tabulated Data - Continued

DATA	MODEL				
	M1008 M1008A1	M1009	M1010	M1028 M1028A1	M1031
~Conitinued)					
Ground Clearance (to T/C skidplate @ GVWR	10.7 in (27.2 cm)	9.6 in (24.4 cm)	10.5 in (26.7 cm)	10.5 in (26.7 cm)	10.5 in (26,7 cm)
Rear Axle @ GVWR					
Front Axle @ GVWR					
Height[overall] @ Curb	7.8 in (19.8 cm)	8.2 in (20.8 cm)	7.7 in (19,6 cm)	7,7 in (19.6 cm)	7,7 in (19.6 cm)
Length[overall]					
Width[overall]					
Engine:	8.6 in (21.8 cm)	8.4 in (21.3 cm)	8.6 in (21.8 cm)	8.6 in (21.8 cm)	8.6 in (21.8 cm)
Type					
No. of Cylinders					
Piston Displacement					
Horsepower					
Fuel					
Allowable Speeds:					
Capacity:	75.9 in (192.8 cm)	74.9 in (190.2 cm)	101.6 in (258 cm)	107,1 in. (272 cm)	76.3 in (193.8 cm)
Fuel Tank					
Tires:					
Size					
Inflation Pressures:					
Maximum Load Capacity:	220.7 in (560.6 cm)	191.8in (487.1 cm)	227.7 in (578.4 cm)	220,7 in (560.6 cm)	212.9 in (504.8 cm)
Front					
Rear	81.2in (206.2 cm)	79.6 in (202 cm)	81.2in (206.2 cm)	81.2in (206.2 cm)	81.2in (206.2 cm)
		V-type		V-type	V-type

1-9.1. DIFFERENCES BETWEEN MODELS

		M1028A2	M1028A3
a.	6.2 Liter Diesel Engine	x	x
b.	Automatic Transmission	x	x
			x
c.	Model 208 Transfer Case (T/C)	x	
d.	Model 205 Transfer Case (T/C)	x	x
e.	100 amp/28 volt Electrical System		
f.	200 amp/28 volt Electrical System	x	x
		x	x
g.	Troop Seat Kit	x	x
		x	x
h.	Communications Kit	x	x
i.	Slave Receptacle		
j.	Tow Pintle		
k.	Cargo Tie-downs	x	x
l.	Shelter Tie-downs		
m.	Air Conditioner		
n. Gas-Particulate Filter Systemo. Locking Differentialp. No-Spki Differentialq. Cargo Cover Kitr. Rear Passenger Seating 8, Spotlightt. Floodlights			

Table 1-1.1 Tabulated Data

DATA	M1028A2 M1028A3	DATA	M1028A2 M1028A3
Make:	Chevrolet	Wheelbase	131.5 in (334 cm)
Weights:		Track (front)	67.8 in (172 cm)
Curb	6120 lb (2778 kg)		
Payload/Pasaengers	3940 lb (1789 kg)	Track (rear)	75.8 In (193 cm)
QVWR	10120 lb max (4594 kg)	Ground Clearance (to T/C skid plate @GVWR)	10.7 in (27.2 cm)
GAWR (front)	3900 lb (1770 kg)	Rear Axle @ GVWR	7.7 in (19.6 cm)
GAWR (rear)	6220 lb (2824 kg)	Front Axle @ GVWR	8.6 in (21.8)

1-9.2. TABULATED DATA

Table 1-1.1 Tabulated Data - Continued

DATA	M1028A2 M1028A3	DATA	M1028A2 M1028A3
Height (overall) @Curb	107,1 in l (272 cm)	Allowable Speeds:	(See Paragraph" 1-13.)
Length (overall)	220.7 in (560.6 cm)	Capacity:	
		Fuel Tank	20 gal (75,7 1)
Width (overall)	95.8 in (243.3 cm)	Tires:	
Engine:		Size	LT235/85R-16E
Type	V-type	Inflation Pressures	(See PMCS, Table 2-1)
No. of Cylinders	8		
Piston Displacement	379 cu in (6.2 1)	Maximum Load Capacity:	
		Front	4500 lb (2043 kg)
Horsepower	135 bhp at 3600 rpm	Rear	7500 lb (3405 kg)
Fuel	Diesel		

* With the S-250 shelter installed. Height will vary with different shelters and loading,

Table 1-1.2 Mission Profiles

	MI 009	M1008, M1008A1 M1028, M1028A1, M1028A2, M1028A3 M1031	M1010
a. Operational Day	20 hours	20 hours	20 hours
b. Usage			
(1) Trips per Day	4	5	6
(2) Miles (km) per Trip	25 ml (40 km)	16 ml (25 km)	16 ml (25 km)
(3) Miles (km) per Day	99 ml (160 km)	78 ml (125 km)	93 ml (150 km)
Road Usage	20%	20%	20%
c. (1) Paved Roads	50%	50%	50%
(2) Secondary Roads	15%	15%	15%
(3) Trails	15%	15%	15%
(4) Cross-country	25%	25%	25%
d. Night Operations			
e. Speed	37 mph (60 kph)	37 mph (60 kph)	50 mph (80 kph)
(1) Day	25 mph (40 kph)	25 mph (40 kph)	19 mph (30 kph)
(a) Paved Road			
(b) Secondary Road			

Table 1-1.2 Mission Profiles - Continued

	MI 009	M1008, M1008A1 M1028, M1028A1, M1028A2, M1028A3 M1031	M1010
(c) Trails	16 mph (25 kph)	16 mph (25 kph)	16 mph (20 kph)
(d) Cross-country	6 mph (10 kph)	6 mph (10 kph)	6 mph (10 kph)
(2) Night (with headlights)			
(a) Paved Road	25 mph (40 kph)	25 mph (40 kph)	25 mph (40 kph)
(b) Secondary Road	25 mph (40 kph)	25 mph (40 kph)	25 mph (40 kph)
(c) Trails	5 mph (8 kph)	5 mph (8 kph)	5 mph (8 kph)
(d) Cross-country	5 mph (8 kph)	5 mph (8 kph)	5 mph (8 kph)
(3) Night (blackout)			
(a) Paved Road	10 mph (16 kph)	10 mph (16 kph)	10 mph (16 kph)
(b) Secondary Road	10 mph (16 kph)	10 mph (16 kph)	10 mph (16 kph)
(c) Trails	4 mph (6 kph)	4 mph (6 kph)	4 mph (6 kph)
(d) Cross-country	2.5 mph (4 kph)	2.5 mph (4 kph)	2.5 mph (4 kph)
f. Fording Depth	16in (40 cm)	16in (40 cm)	16in (40 cm)

TM 9-2320-289-10

Section III. TECHNICAL PRINCIPLES OF OPERATION

1-10. ENGINE

a. The CUCV Series of trucks are powered by a V-8 diesel engine which has a displacement of 6.2 liters. This engine is similar to a V-8 gasoline engine in many ways, but major differences occur in cylinder heads, combustion chambers, fuel distribution system, air intake manifold, and method of ignition.

(1) Fuel Distribution: There is no carburetor on a diesel engine. Instead, fuel is injected under high pressure through nozzles into the combustion chamber at correct timing intervals.

(2) Ignition. Ignition of fuel occurs because of heat developed in combustion chamber. As a result, no spark plugs or high voltage ignition system is required. For cold starts, glow plugs heat up combustion chamber.

(3) Air Intake Manifold. Air intake manifold is always open to atmospheric pressure. As a result, the engine does not develop a vacuum supply. Therefore, the engine drives a vacuum pump which supplies shift signals to the automatic transmission.

b. The lower engine (case, crankshaft, camshaft, bearings, rods, pistons, and wrist pins) is similar to a gasoline engine, but is of a heavy-duty design because of greater stresses developed in a diesel engine.

c. Here are some driving tips that will help you maintain good engine performance:

(1) DO NOT make full-throttle starts and hard stops.

(2) Take it easy when engine is cold. Maximum performance is reached at normal operating temperatures.

(3) DO NOT hold truck on an uphill grade with accelerator pedal. Use the regular brakes to hold truck.

(4) DO NOT use starter aids such as ether or gasoline in air intake system. Such aids can cause immediate engine damage.

(5) Pumping accelerator pedal before or during cranking will not help start engine.

(6) When engine is cold, let it Idle for a few seconds before driving; this will allow proper oil pressure to build up. (Increased operating noise and light exhaust smoke are normal when engine is cold.)

(7) If oil pressure light comes on while operating truck, IMMEDIATELY SHUT DOWN ENGINE. Continued operation of engine while it is not being properly lubricated will cause serious damage.

(8) DO NOT use diesel fuel which has been contaminated with engine oils.

(9) Maintain a minimum of 1/4 tank of fuel when temperatures are below 20°F.

(lo) DO NOT run engine at idle for extended periods of time; damage can occur to engine.

1-11. FUEL SYSTEM

a. A combination of low pressure and high pressure pumping systems moves fuel from the fuel tank to the engine.

b. Fuel is pumped from the fuel tank by a low pressure pump which moves fuel toward the fuel filter that separates water and contaminants from the fuel. From the fuel filter, fuel is drawn into the high pressure injection pump which meters and pressurizes fuel and sends it through high pressure fuel lines into the injection nozzle located in the engine's precombustion chamber.

1-13

TM 9-2320-289-10

1-12. TRANSMISSION

    a. All CUCV trucks have an automatic transmission with three forward and one reverse speeds. The transmission provides the driver with a selection of vehicle speeds while holding engine speeds within effective torque range. It also allows disengaging and reversing flow of power from engine to wheels. Automatic shifting schedules are controlled by a vacuum modulator that allows for smooth shifts by "sensing" engine load changes.

    b. The transmission gearshift lever(l), located on the steering column, provides the following selections on the gearshift indicator (2):

        (1) "P" (Park). Transmission locked; truck will not move.

        (2) "R" (Reverse). Puts truck in reverse for backing operations.

        (3) "N" (Neutral). Transmission mechanism disengaged; truck wheels can move by coasting but truck is not in gear.

        (4) "D" (Drive). For normal driving with light to moderate speeds; automatic downshift at speeds under 35 mph (56 kph) by depressing accelerator about halfway to the floor; forced downshift at speeds above 35 mph (56 kph) by depressing pedal all the way to the floor.

        (5) "2." For hill climbing or engine braking to slow truck when going down a steep hill. Gearshift lever maybe shifted from "D" to "2" and from "2" to "D" under most driving conditions.

        (6) "1" For maximum engine braking when driving down very steep hills or when maximum performance is required to climb a very steep hill or drive through deep snow or mud. You may shift into "l" at any speed, but the transmission will not lock in "l" until your truck's speed is below 40 mph (64 kph).

c. Here are some driving tips which will help protect the transmission from damage:

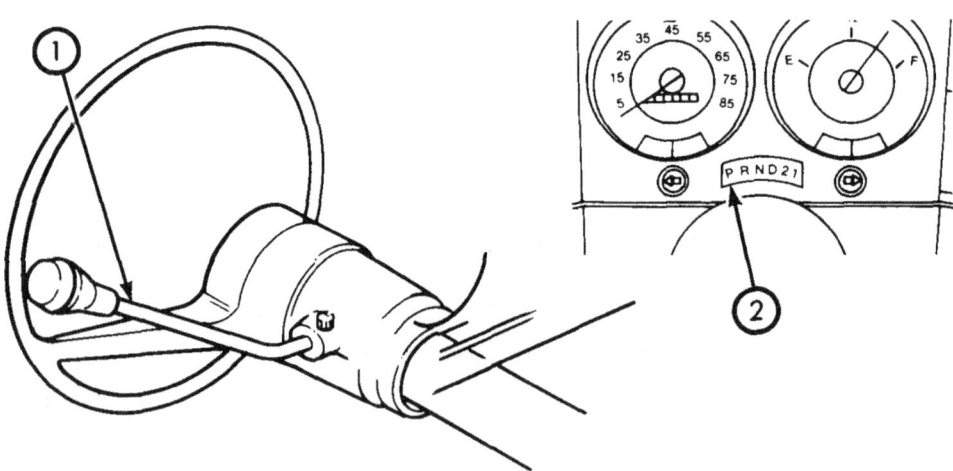

    (1) DO NOT coast downhill in "N" (Neutral).

    (2) NEVER shift transmission gearshift lever to "P" (Park) or "N" (Neutral) while truck is in motion.

TA466572

TM 9-2320-289-10

(3) DO NOT race engine when shifting from "P" (Park) or "N" (Neutral) into another gear range.

(4) DO NOT shift between forward ranges and "R" (Reverse) while operating engine at high speed or heavy throttle.

(5) DO NOT force transmission gearshift lever.

(6) DO NOT operate transmission at or near "stall condition" for more than 10 seconds at a time. ("'Stall condition" is when the engine is running at high speed while the transmission is in a driving range and drive wheels aren't moving, such as when the truck is stuck in deep sand, mud, deep snow, or when truck is against a fixed barrier.)

(7) DO NOT shift transmission gearshift lever to "P" (Park) on a hill before setting parking brake. This puts force on the transmission and makes it difficult to shift the transmission gearshift lever out of "P" (Park). Make sure the transfer case is in gear.

(8) When preparing to drive, DO NOT release parking brake until transmission gearshift lever is shifted out of (Park) position.

1-13. TRANSFER CASE

　　a. All trucks, except the M1028A1, M1028A2, and M1031, are equipped with a New Process Model 208 transfer case. The M1028A1, M1028A2, and M1031 are equipped with New Process Model 205 transfer case which is designed to accommodate installation of a power take-off unit. (A power take-off unit uses engine power to run equipment.)

　　b. The transfer case transmits engine power to front axle for four-wheel drive operation. Four-wheel drive is used to provide additional traction and lower gearing for use in off-road operations and to provide low speed pulling power in unusual conditions.

　　c. The four transfer case control lever positions are:

　　　(1) "N" (Neutral). In this position, both front and rear axles are not engaged. Power from engine will not turn wheels.

　　　(2) "2H" (Two-wheel Drive High Range). This position is used for normal, two-wheel driving on dry, primary and secondary roads at normal highway driving speeds.

　　　(3) "4L" (Four-wheel Drive Low Range). This position is used for driving under unusual conditions in LOW speed ranges.

　　　(4) "4H" (Four-wheel Drive High Range). This position is used for driving under unusual conditions in HIGH speed ranges.

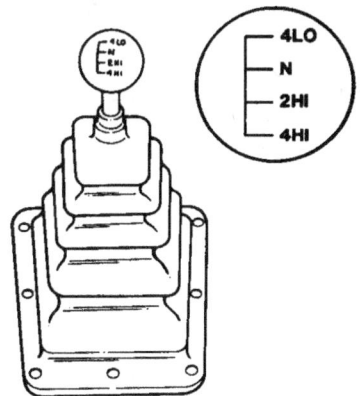 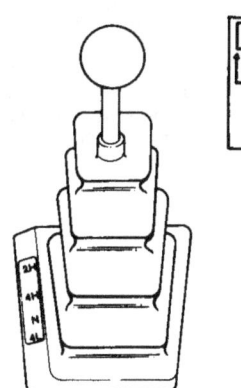

M1028A1, M1028A2, and M1031　　All Other

TA466573

Change 3 1-15

TM 9-2320-289-10

    d. Four-wheel drive ranges of transfer case should be used only when greater traction and power are required in off-road operations. Thus, transfer case control lever can be left in "2H" most of the time.

    e. Here are some driving tips which will help protect transfer case from damage:

        (1) DO NOT use "4L" or "4H" transfer case ranges on dry, hard-surfaced roads, or premature wear or excessive damage to transfer case and/or tires may result.

        (2) DO NOT complete transfer case shift before ensuring that both locking hubs are in the same position.

        (3) DO NOT use tools to assist in turning locking hubs,

        (4) DO NOT exceed the following maximum speeds when transfer case gears are engaged:

	Transfer Case In Four-high ("4H")	Transfer Case In Two-high ("2 H")	Transfer Case In Four-low ("4L")
Trans Gear			
1	25 mph (40.2 kph)	14 mph (22.5 kph)	25 mph (40.2 kph)
2	35 mph (56.3 kph)	23 mph (37,0 kph)	35 mph (56.3 kph)
Drive	55 mph (88.5 kph)	35 mph (56.3 kph)	55 mph (88.5 kph)
Reverse	9 mph (24.5 kph)	6 mph (9.6 kph)	9 mph (14,5 kph)

        (5) The truck's speedometer does not indicate reverse speeds. You must estimate your speed and use good judgement when driving in reverse.

        (6) The ideal gear to use for backing operations is the "2H" position. However, if the truck is in "4L" or "4H" and becomes mired in deep mud or snow, shifting to "2H'" at this time will dig the truck in deeper. Leave truck in four-wheel "drive position, but take care not to exceed "4L" or "4H" speed limits.

1-14. ELECTRICAL SYSTEM

    a. The CUCV electrical system consists of the following major components:

        (1) Battery System. Consists of two 12 volt batteries in series which supply 24 volts to starting, charging, and glow plug systems, and 12 volts to wiring and lighting system.

        (2) Charging System. Consists of two alternators, voltage regulators, and instrument panel indicators.

        (3) Starting System. Includes ignition switch, starter relay, solenoid, and starter motor, Transmission gearshift lever is mechanically linked to the ignition. The starter safety ignition switch is designed such that the truck will only start when the transmission gearshift lever is shifted to either "P" (Park) or "N" (Neutral) position,

TM 9-2320-289-10

(4) Glow Plug System. Consists of 8 glow plugs (one for each cylinder) which are used to preheat combustion chamber for easier starting. There is a WAIT (glow plug indicator) light (1) on the control panel which lights up to tell you that the glow plugs are in the process of warming the combustion chamber. When combustion chambers are warmed to starting temperatures, the WAIT light will go out.

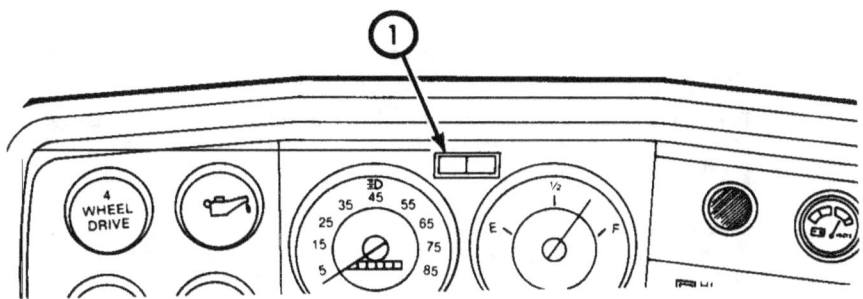

(5) Wiring and Lighting System. Includes wiring harnesses, fuses, sockets, and lamps.

b. Here are some important tips about the electrical system that will help you properly operate the truck:

(1) If you leave ignition in "ON," but do not start engine, the glow plugs will continue to operate and could drain batteries.

(2) If engine is warm, the WAIT light (1) may not come on, or it may go on and off a few times during cranking or after starting. This is normal.

(3) If WAIT light (1) stays on, fails to go out, or comes back on after engine is warmed UP, there may be a system malfunction and you should immediately notify your supervisor.

(4) If generator lights do not go out within a few seconds, press accelerator and let engine return to idle; lights should go out. If not, notify your supervisor.

(5) The light switch controls headlights, taillights, parking lights, and blackout lights.

(6) Windshield wipers will not work when ignition switch is off.

(7) The service lights/blackout toggle switch on instrument panel must be in "SERVICE LIGHTS" position, otherwise brake lights, hazard warning lights, and turn signals on truck will not work.

(8) Turn signals do not work when hazard warning flasher is on.

(9) If service brake pedal is depressed, hazard warning flasher will not flash.

(lo) Hazard warning flasher will work with ignition turned on or off.

1-15. BRAKE SYSTEM

a. The CUCV braking system uses hydraulically activated, power assisted front disc and rear drum service brakes. Fluid for service brakes is contained in the master cylinder. The master cylinder operates through hydraulic energy supplied by the power steering pump and power booster.

TM 9-2320-289-10

b. The following is important safety information about the CUCV'S brake system:

(1) If power assist is lost, brakes normally can be applied with power assist at least two times using reserve power. Without power assist, the truck can still be stopped by depressing much harder on service brake pedal, but stopping distance will be longer.

(2) Do not pump service brakes when brake power assist has been lost. Pumping brakes will use up reserve fluid.

(3) Riding brake by resting your foot on service brake pedal can cause overheating of brakes. This can cause unnecessary brake wear, as well as waste fuel.

(4) To prevent premature brake wear, make sure parking brake is fully released. The parking brake warning light is designed to stay on if parking brake is not fully released and ignition key is in "ON" position.

1-16. STEERING SYSTEM

The steering system includes steering wheel and column, steering gear, and steering linkage. Steering is power assisted by fluid that is pumped by the power steering pump, which sends fluid first to the power booster unit of the brake system. From there, fluid goes to the steering gear.

1-17. AMBULANCE PECULIAR COMPONENTS

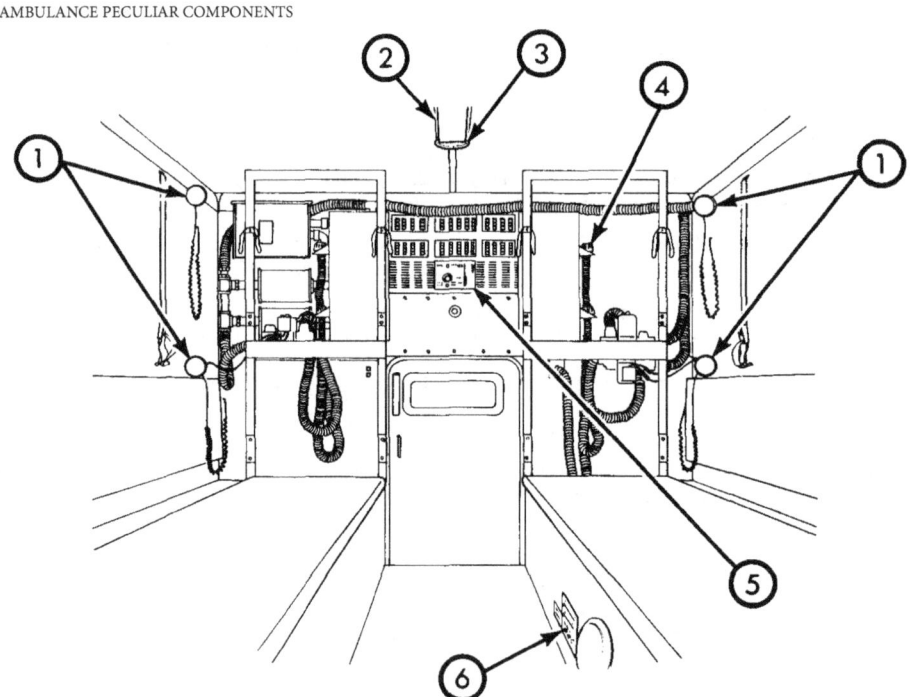

The CUCV M 1010 Ambulance contains a number of components which are not found on the other trucks. Some of these components include:

a. Gas-Particulate Filter Unit (GPFU). The GPFU provides the ambulance with clean, filtered, breathable air that is free of chemical agents and is heated to a comfortable temperature. Air is channeled to seven hookups (4), six of which can be used simultaneously. Two of the hookups are in the cab; the remainder are in the patient compartment. The GPFU is designed to function with M25 Series protective masks. The filter unit and masks WILL NOT protect against carbon monoxide,

TA466575

1-18

TM 9-2320-289-10

b. Air Conditioning Unit The air conditioning unit (5) cools end ventilates the ambulance patient compartment. It is mounted on the front wall of the patient compartment, and is designed to provide a section between cooling and recirculating inside air and cooling outside air. The inside air section is used when maximum cooling is required under conditions of high temperature and humidity. The outside air is used for most air conditioning situations and for vent mode. Fan speed end temperature may be varied as required.

c. Personnal Heater. Because the heater in the cab is not sufficient to heat the patient compartment, e personnel heater is provided within the patiant compartment. It produces heat by burning a mixture of fuel and air in a stainless steel heat exchanger and burner assembly. The heater is designed to bum fuel from the trucks fuel tank. It can be manually controlled from a control unit
(6) located at the base of the lower litter berth on the right side of the patient compartment.

d. Domelight and Focus Lights. There is a fluorescent domelight (2) over the canter aisle of the patient compartment to provide light for medical purposes. When the service lights/blackout toggle switch the "'BLACKOUT" position and the rear doors or door between the cab and patient compartment is open. the domelight goes off and a blackout light (3) comes on. When more light is needed, four focus lights (1) (two on each wall) can be used. The lights can be pulled from thair wall mounts and mounted in any of six mounting blocks, or they can be hand held if needed.

1-18. TRUCK LOADING

a. Proper loading of the truck is governed by weight limits and load distribution. The trucks era rated in tame of Gross Vehicle Weight Rating (GVWR) and Gross Axle Weight Rating (GAWR). These ratings are beead upon the weekest component in the suspension system (axles, springs, tires, or wheels.

(1) Gross Vehicle Weight Rating (GVWR). The Gross Vehicle Waight Rating (GVWR) is the maximum allowable loaded weight of the truck, driver, passenger(s), and payload, measured in pounds. Tha rating takes into account the capabilities of the engine, trarsmission, frame, springs, brakes, axles, and tires.

(2) Gross Axle Welght Rating (GAWR). The maximum load that can be put on front and rear axles.

b. A certification label, located on the trailing edge of the operator's left-hand door, specifies these ratings and shows the maximum waight that the front axle can carry (front GAWR) and tha maximum weight that the rear axle can carry (rear GAWR).

c. The example below shows a fully loaded truck, tha maximum GVWR (1), tha front GAWR (3), and the back GAWR (2).
Curb weight equals the weight of the truck without driver, passenger or cargo, but includes fuel and coolant.

EXAMPLE ONLY

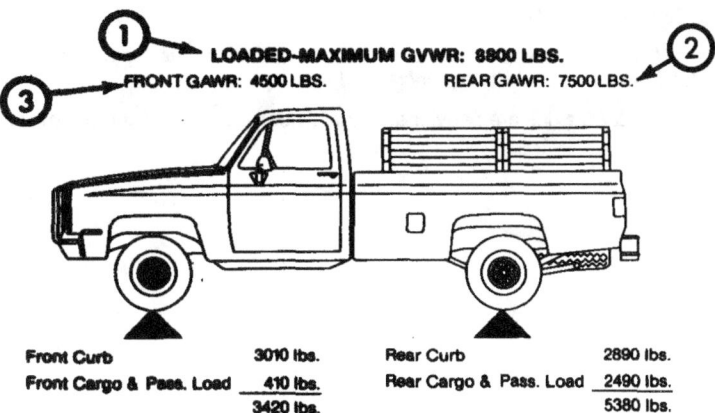

TA466576

Change 5   1-19

TM 9-2320-289-10

d. Follow these guidelines for proper truck loading:

<u>WARNING</u>

- This vehicle has been designed to operate safely and efficiently within the limits specified in this TM. Operation beyond these limits is prohibited IAW 70-1 without written approval from the Commander, U.S. Army Tank-Automotive Command, ATTN: AMSTA-CM-S, Warren MI 48397-5000.

- DO NOT EXCEED GVWR and GAWR. overloading in of truck control and personnel injury, either by causing component failure or by affect-ing the truck's handling. In addition overloading can cause serious damage to the truck's suspension system.

(1) NEVER exceed GVWR and GAWR.

(2) The cargo load should be placed on both sides of the centerline marked on the shipping plate as equally as possible.

(3) Always put load to the front of the cargo box first.

(4) Ensure that the tires on your truck are properly inflated for the load which you are carrying.

1-19. WINTERIZATION KIT

a. CUCV trucks operating under arctic conditions will be equipped with winterization equipment. This equipment consists of an engine coolant and oil heater, a personnel heater (for ambulance and those trucks with an enclosed cargo area), and a battery and passenger compartment heater.

b. The 24 v personnel heater is installed on the right front corner of the cargo area floor. (In the ambulance, this heater is located under the rear of the right bench.) This is the same multifuel combustion type heater described in the ambulance section above.

c. The battery and passenger compartment heater is located under the hood on the right fender. This heater is the same type as the personnel heater.

d. The engine coolant heater, located under the hood on the left fender, is designed to preheat diesel engines for starting at temperatures down to -65°F. It operates on 24 volts and burns liquid fuel. It has a built-in pump to circulate the heated coolant in the engine's coolant system.

1-20. DUAL WHEEL CONFIGURATION

a. M1028A2 and M1028A3 are shelter carriers that have been converted to dual rear wheels. This provides greater load carrying capability and increased stability.

b. M1028A2 and M1028A3 axles are heavier and dual wheels spread loads more evenly.

TM 9-2320-289-10

CHAPTER 2

OPERATING PROCEDURES

Section I. DESCRIPTION AND USE OF CONTROLS AND
INDICATORS

2-1. INTRODUCTION

The information and illustrations in this chapter provide the basic instructions you will need in order to properly operate the truck. Before operating the truck, make sure you know the location and operation of all controls and instruments. A thorough review of this section is the best way to do this. Get into the cab of the truck and identify each control and instrument as you come upon it in this section.

2-2. CONTROLS ANO INDICATORS

The figures below show the truck's instrument panel. Very often the lights and gages indicate that something is wrong long before you realize it. Know them before you operate the truck.

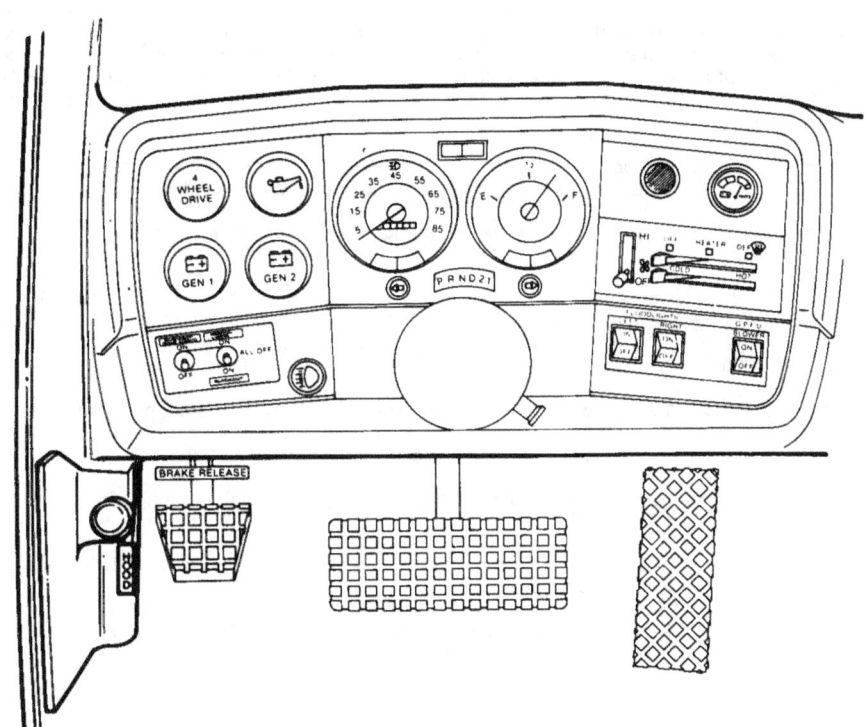

TM 9-2320-289-10

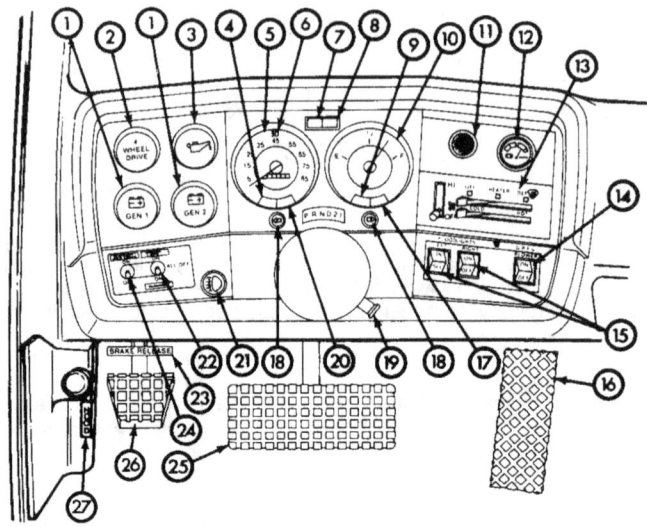

1. **GEN 1 and GEN 2 Lights.** On all models except the M1010 Ambulance, generator lights come on when ignition key is in the "ON" position, but before engine is started. After engine starts, lights should go out and remain out. If either or both lights remain on after engine starts, press accelerator and let engine return to idle. If lights remain on or come on during operation of truck, IMMEDIATELY SHUT DOWN ENGINE and notify your supervisor. M1010 Ambulance has no generator warning lights.

2. **Four-wheel Drive Indicator Light.** Indicates when transfer is in "4L" or "4H" position. Light will remain on until the transfer case control lever is shifted to either "N" (Neutral) or "H."

3. **Oil Pressure Light.** Indicates lack of oil pressure being delivered to parts of engine requiring lubrication.

4. **Engine Coolant Temperature Light.** Indicates that overheating exists when light comes on. If overheating is indicated, operator shall take action by following steps in Troubleshoot-ing, Table 3-1. Make a practice of observing this light and others while driving, especially in hot weather and when truck has a heavy load.

5. **Speedometer.** Indicates truck's speed in miles per hour (mph) and kilometers per hour (kph).

6. **Headlight High Beam Indicator Light.** Indicates when headlights are on high beam.

7. **WAIT (Glow Plug Indicator) Light.** Indicates when glow plug system is operating. For further information, refer to paragraph 1-14.

8. **WATER-IN-FUEL Indicator Light.** Indicates when there is water in fuel system. When water is present in fuel system, drain it. (See paragraph 3-7)

9. **Seat Belt Indicator Light.** Lights up during start-up (for about 4 to 8 seconds). If operator is not wearing a seat belt, buzzer will sound for 4 to 8 seconds.

1. **Fuel Gage.** Indicates amount of fuel in fuel tank.

2. **Door Ajar Indicator Light (M1010 Ambulance Only).** Indicates that back doors of patient compartment are not latched closed. Make a practice of observing this light when driving.

3. **Voltmeter.** Indicates electrical system voltage when engine is running. When voltmeter indicator is in green zone, charging system is operating correctly. When indicator is in yellow zone or left-hand red zone, system is not charging. When indicator is in right-hand red zone, system is overcharging.

TA466578

2-2

TM 9-2320-289-10

13. Heater/Defroster Controls. Selects heater or defroster mode, controls temperature of heater and defroster, and controls speed of fan.

14. Gas-Particulate Filter Unit Controls (M1010 Ambulance Only). "ON" and "OFF" controls for gas-particulate filter unit in patient compartment. (See paragraph 2-25 for details on unit)

15. Floodlight Controls (M1010 Ambulance Only). "ON" and "OFF" controls for two floodlights which are mounted on the outside of ambulance.

16. Accelerator Padal Depress pedal to increase engine speed.

17. Brake System Warning Light. Indicates if there is a malfunction in brake system. It also stays on when parking brake is set or not fully released and ignition key is in "ON" position. It comes on briefly during engine start-up so that you can check that bulb is working. If light does not come on when starting or when parking brake is set, system should be repaired.

Turn Signal Indicators. When turn signal lever is pushed up, right light flashes. When turn signal lever is pushed down, left light flashes. When hazard warning button is pushed, both lights flash.

18. Hazard Warning Flasher. Push in on button to turn on hazard warning flasher. Pull out button to turn off hazard warning flashers.

19.

20. Low Coolant Warning Light. Lights up during start-up of engine to serve as a bulb check. If light does not go out after engine starts or it comes on while driving, check radiator coolant level.

21. Light Switch. A pullout knob that turns on service lights, blackout markers, and blackout drive light when switches (22) and (24) are in appropriate positions. It also controls brightness of instrument panel lights. Brightness can be changed by turning knob clockwise or counterclockwise.

22. Servica Lights/Blackout Toggle Switch. This toggle switch must be moved to "SERVICE LIGHTS ON" before brake lights, hazard warning lights, or horn will operate. It must also be used in conjunction with light switch (21) to turn on service lights or blackout markers. Refer to paragraph 2-15 for operating instructions. Service lights/blackout toggle switch should remain in "OFF" position when truck is not in use.

23. Braka Release Handle. When pulled out, it releases parking brake.

24. Blackout Drive Switch. This toggle switch must be moved to "BLACKOUT DRIVE LIGHT ON" and released for blackout drive to work. This is a self-centering toggle and will not stay in "ON" position. Refer to paragraph 2-15 for operating instructions.

25. Service Brake Pedal. Depress pedal to apply service brakes to stop truck.

26. Parking Brake Pedal. Depress pedal to set parking brake.

26. Hood Handle. Pull this handle to release hood so that it can be opened.

2-3. PREPARATION FOR USE

27. All newly received trucks must be road tested to check their operation and general condition. This test can be performed during your normal duties. The test must be long enough to allow complete observation of trucks operating condition.

b. While operating truck, observe instrument panel lights and gages for any indication of faulty truck operations. Stop truck and discontinue test if any serious trouble develops. Pay special attention during road test to oil pressure and engine coolant temperature lights, WATER-IN-FUEL indicator, and low coolant warning light. If you see any indication of faulty operation, IMMEDIATELY SHUT DOWN ENGINE. Notify your supervisor and DO NOT attempt to operate truck any further.

2-3

TM 9-2320-289-10

WARNING

NEVER remove radiator cap when engine is hot. This is a pressurized cooling system and escaping steam or hot water can cause serious burns.

c. Stop at least twice in first 25 miles (40km) to check for leaking coolant, oil, fuel, or audible exhaust noise. Discontinue test driving if you find any of these troubles.

d. Report any controls that are hard to operate or instruments that give irregular readings. Be alert for unusual noises or vibrations.

e. After completing or discontinuing driving test, report any problems to your supervisor,

TM 9-2320-289-10

Section II. PREVENTIVE MAINTENANCE CHECKS AND SERVICES (PMCS)

2-4. GENERAL

Preventive Maintenance Checks and Services (PMCS) means systematic caring, inspecting, and servicing of equipment to keep it in good condition and to prevent breakdowns. As the truck's operator, your mission is to:

a. Be sure to perform your PMCS each time you operate the truck. Always do your PMCS in the same order, so it gets to be a habit. Once you've had some practice, you'll quickly spot anything wrong.

b. Do your BEFORE (B) PMCS just before you operate the truck. Pay attention to WARNINGS, CAUTIONS, and NOTES.

c. Do your DURING (D) PMCS while you operate the truck. During operation means to monitor the truck and its related components while it is actually being operated. Pay attention to WARNINGS, CAUTIONS, and NOTES.

d. Do your AFTER (A) PMCS right after operating the truck. Pay attention to WARNINGS, CAUTIONS, and NOTES.

e. Do your WEEKLY (W) PMCS once a week.

f. Do your MONTHLY (M) PMCS once a month.

g. Use DA Form 2404 (Equipment Inspection and Maintenance Worksheet) to record any faults that you discover before, during, or after operation, unless you can fix them. You DO NOT need to record faults that you fix.

h. Be prepared to assist organizational maintenance when they lubricate the truck. Perform any other services when required by organizational maintenance.

2-5. PMCS PROCEDURES

a. Your Preventive Maintenance Checks and Services, Table 2-1, lists inspections and care required to keep your truck in good operating condition. It is set up so you can make your BEFORE (B) OPERATION checks as you walk around the truck.

b. The "INTERVAL" column of Table 2-1 tells you when to do a certain check or service.

c. The "PROCEDURE" column of Table 2-1 tells you how to do required checks and services. Carefully follow these instructions. If you do not have tools, or if the procedure tells you to, notify your supervisor.

NOTE

Terms "ready/available" and "mission capable" refer to sama status: Equipment is on hsnd and ready to parform its combat missions. (Sac DA Pam 738-750)

d. The "EQUIPMENT IS NOT READY/AVAILABLE IF:" column in Table 2-1 tells you when your truck is nonmission capable and why the truck cannot be used.

e. If the truck does not perform as required, refer to Chapter 3, Section II, Troubleshooting.

f. If anything looks wrong and you can't fix it, write it on your DA Form 2404. IMMEDIATELY, report it to your supervisor.

TM 9-2320-289-10

g. When you do your PMCS, you will always need a rag or two. Following are checks that are common to the entire truck:

(1) Keep It Clean. Dirt, grease, oil, and debris only get in the way and may cover up a serious problem. Clean as you work and as needed, Use dry cleaning solvent (SD-2) on all metal surfaces. Use soap and water when you clean rubber or plastic material, Upholstery can be cleaned with soap and water and a clean, damp cloth,

(2) Rust and Corrosion. Check truck body and frame for rust and corrosion. If any bare metal or corrosion exists, clean, and apply a thin coat of oil. Report it to your supervisor.

(3) Bolts, Nuts, and Screws. Check them all for obvious looseness, missing, bent, or broken condition. You can't try them all with a tool, but look for chipped paint, bare metal, or rust around bolt heads. If you find a bolt, nut, or screw you think is loose, tighten it or report it to your supervisor.

(4) Welds. Look for loose or chipped paint, rust, or gaps where parts are welded together, If you find a bad weld, report it to your supervisor,

(5) Electric Wires and Connectors. Look for cracked, frayed, or broken insulation, bare wires, and loose or broken connectors. Tighten loose connectors. Report any damaged wires to your supervisor.

(6) Hoses and Fluid Lines. Look for wear, damage, and leaks, and make sure clamps and fittings are tight. Wet spots show leaks, but a stain around a fitting or connector can also mean a leak. If a leak comes from a loose fitting or connector, tighten it, If something is broken or worn out, report it to your supervisor.

h. When you check for "operating condition," you look at the component to see if it's serviceable.

2-6. CLEANING AGENTS

WARNING

DO NOT use diesel fuel, gasoline, or benzene (benzol) for cleaning.

- DO NOT SMOKE when using cleaning solvent. NEVER USE IT NEAR AN OPEN FLAME. Be sure there is a fire extinguisher nearby and use cleaning solvent only in well-ventilated places. Flash point of solvent is 138°F (60°C).

- USE CAUTION when using cleaning solvents. Cleaning solvents evaporate quickly and can irritata axposed skin if solvents contact skin. In cold weather, contact of exposad skin with cleaning solvents can cause frostbite.

CAUTION

When cleaning undarhood areas, angina must be COLD (same temperature as outside air). DO NOT point water or steam directly at any electrical connection. DO NOT point water stream directly at radiator fins. DO NOT use high pressura water supply systam. Damage to engine, electrical system, and other compo-nents may result.

NOTE

Only use those authorized cleaning solvents or agents listed in Appendix D.

Cleaning Underhood Areas.

a (1) When using water to clean the engine compartment, always cover alternators and air cleaner inlet using waterproof material. For M1010 only, also cover air conditioner compressor. Use water pressure and volume similar to a standard household type water supply system (45-70psi, 6.5-10.2 kPa).

(2) After cleaning, allow engine to air dry. Do not use compressed air to dry engine. Do not run engine to decrease drying time.

(3) Remove all component covers before starting engine.

b. Treating Mildewed Areas. Canvas that has mildewed can be cleaned by scrubbing with a dry brush. If it is necessary to use water to remove dirt, it should not be used until mildew has been removed. After removing mildew, examine fabric. Look for evidence of deterioration. If canvas has deteriorated, it should be replaced.

<u>CAUTION</u>

Keep cleaning solvents, gasoline, and lubricants away from rubber or soft plastic parts. They will deteriorate material.

c. Cleaning Rust or Grease. When cleaning grease buildup or rusty places, use a cleaning solvent. Then apply a thin coat of light oil to affected area.

2-7. LEAKAGE DEFINITIONS FOR OPERATOR PMCS

It is necessary for you to know how fluid leakage affects the status of the truck. Following are type/classes of leakage an operator needs to know to be able to determine the status of the truck. Learn these leakage definitions and remember - when in doubt, notify your supervisor.

<u>CAUTION</u>

Equipment operation is allowable with minor leakages (Class I or II). Of course, consideration must be given to fluid capacity in the item/system being checked/inspected. When in doubt, notify your suPervisor.

When operating with Class I or II leaks, continue to check fluid levels as required in your PMCS.

Class III leaks should be reported immediately to your supervisor.

a. CLASS I - Seepage of fluid (as indicated by wetness or discoloration) not great enough to form drops.

b. CLASS II - Leakage of fluid great enough to form drops but not enough to cause drops to drip from item being checked/inspected.

c. CLASS III - Leakage of fluid great enough to form drops that fall from item being checked/inspected.

TM 9-2320-289-10

Table 2-1. Preventive Maintenance Checks and Services for Model CUCV

Item No.	Interval	Location Item to ChecK Service	Crewmember Procedure	Not Fully Mission Capable If:
			WARNING Always Remember The WARNINGS, CAUTIONS And NOTES Before Operating This Vehicle And Prior to PMCS. Perform all PMCS checks if: a. You are the assigned driver but have not oper-ated the vehicle since the last weekly inspection. b. You are oper-ating the vehicle for the first time. c. See separate manuals for Smoke Generator, Tow Launcher and Radios.	

2-8 Change 4

TM 9-2320-289-10

Table 2-1. Preventive Maintenance Checks and Services Model CUCV

Item No.	Interval	Location / Item to Check/Service	Crewmember Procedure	Not Fully Mission Capable If:
1	Before	Left Front, Side Exterior	NOTE If leakage is detected, further investigation is needed to de-termine the location and cause of the leak. a. Check underneath vehi-cle for evidence of fluid leakage. b. Visually check left side of vehicle for obvious dam-age that would impair op-eration.	a. Class III leak of oil, fuel or coolant. b. Any damage that will prevent opera-tion.
2	Before	Left Side Tires	WARNING Operating a vehicle with a tire in an under-inflated con-dition or with a ques-tionable defect, may lead to prema-ture tire failure and may cause equipment damage, injury or death to personnel. Visually check tire for presence and under-inflation,	Tire missing, deflated or unserviceable.

Change 4 2-9

TM 9-2320-289-10

Table 2-1. Preventive Maintenance Checks and Services for Model CUCV

	Interval	Location Item to Check/ Service	Crewmember Procedure	Not Fully Mission Capable If:
	Before	Rear Exterior	DRIVER Check rear of vehicle for obvi-ous damage that would impair operation.	Any damage that will prevent operation.
	Before	Right Front and Side Exterior	NOTE If leakage is detected, further investigation is needed to de-termine the location and cause of the leak. Check right side of vehicle for obvious damage that would im-pair operation. Visually check tires for pres-ence and under-inflation.	Any damage that will prevent operation.
	Before	Right Side Tires		Tire missing, de-flated or unservice-able.
	Before	Front	NOTE If leakage is detected, investi-gation is needed to determine the location and cause of the leak. a. Check body for visual dam-age that would impair opera-tion of the vehicle. b. Look under vehicle for evi-dence of fluid leakage.	a. Any damage that will prevent opera-tion. b. Class III leak of oil, fuel or coolant.

2-10   Change 4

Table 2-1. Preventive Maintenance Checks and Services Model CUCV

Item No.	Interval	Location / Item to Check Service	Crewmember Procedure	Not Fully Mission Capable If:
7	Before	Seat Belts and Seats	NOTE Vehicle operation with inopera-tive seat belts may violate AR 385-55.a. Check all seat belts for se-curity, damage and complete-ness. NOTE The seat back lock on M1009 vehicles does not activate un-less vehicle is at a 30 degree incline.b. Check operation of seat ad-justing mechanism.	b. Seat adjustmentlock broken or miss-ing.
8	Before	Fire Ex-tin-guisher	a. Check for missing or dam-aged fire extinguisher.	a. Fire extinguishermiss-ing or dam-aged.

Change 4 2-1 1

TM 9-2320-289-10

Table 2-1. Preventive Maintenance Checks and Services for Model CUCV

Item No.	Interval	Location Item to Check/Service	Crewmember Procedure	Not Fully Mission Capable If:
8	Before	Fire Extinguisher Continued	DRIVER b. Check gage for proper pressure of about 150 psi (1034 kPa). c. Check for damaged or missing seal.	b. Pressure gage needle in recharge area. c. Seal broken or missing.
9	Before	Gear Shifter	a. Check transmission shift lever operation. With key in 'ON" position, move shift lever from "P" to "1" then back to "P". Lever should move freely through all range detents. b. Check transfer shift lever operation. With transmission in "N" shift transfer lever through all range positions. Lever should move freely through all range positions.	a. Lever inoperable or binds between range detents. b. Lever inoperable or binds between range detents.

2-12   Change 4

Table 2-1. Preventive Maintenance Checks and Services Model CUCV

Item No.	Interval	Location Item to Check Service	Crewmember Procedure	Not Fully Mission Capable If:
10	Before	Instrument Panel	DRIVER WARNING If gages, instruments, or in-strument lights are not oper-ating as described in thesechecks, IMMEDIATELY shutoff en-gine and notify super-visor or unit main-tenancepersonnel. Continued opera-tion of vehicle may result inpersonnel injury or damageto equipment. NOTE If engine is warm, WAIT lightmay not come on. Duringcranking or after starting, lightmay goon and off a few times.a. Check WAIT light. Turntruck ignition key to the"ON" position. Light shouldcome on if engine is cold. Light will go off when engineis ready to be started.b. Check generator lights (all except M1010). Lightsshould come on when key isin "ON" position and go offafter engine has started.c. Check engine oil light. Light should come on whenkey is in "ON" position andgo off after engine is run-ning.	a. WAIT lightdoes not come onwhen engine iscold, or WAIT lightstays on con-tinously. b. One or bothlights remain onafter engine has started.c. Light does notflash during startup, or remains onafter engine is run-ning.

TM 9-2320-289-10

Table 2-1. Preventive Maintenance Checks and Services for Model CUCV

Item No.	Interval	Location Item to Check/ Service	Crewmember Procedure	Not Fully Mission Capable If:
10	Before	Instrument Panel Continued	DRIVER d. Start engine and watch lights and gages. e. Check door ajar light (M1010 only). Light on indicates patient compartment doors not securely latched. f. Check operation of "ON" "OFF" gas particulate filter unit switch (M1010 only). g. Check engine coolant temperature light. Light should flash momentarily during start up. h. Check low coolant warning light. Light should flash momentarily during start up. i. Check water-in-fuel light. Light should flash momentarily during start up. Drain water from fuel filter immediately if light stays on. j. Check voltmeter. k. Check 4-wheel drive indicator light. With transmission in N, shift transfer to 4H and then to 4L. Light should come on with shifter in these positions. Light should go off when shifter is moved out of 4H or 4L.	d. Engine will not start. g. Light does not flash during startup or remains on when engine is running. h. Light does not flash during startup or remains on when engine is running. j. Voltmeter inoperative or stays in the red or yellow.

2-14 Change 4

Table 2-1. Preventive Maintenance Checks and Services Model CUCV

Item No.	Interval	Location Item to Check/ Service	Crewmember Procedure	Not Fully Mission Capable If:
11	Before	Steering	DRIVER Check steering wheel for op-eration. With engine running, turn steering wheel from left to right. Steering wheel should move freely.	Steering wheel inop.erable or binds.
12	Before	Brake System	a. Check brake system warning light. Light should come on when key is in "ON" position. NOTE Engine must be warmed up and idling, transmission in "D" (drive), transfer in "2N" (two wheel drive) and parking brake released to perform the follow-ing check. b. Check parking brake. Parking brake fully applied, transmis-sion in D or R, transfer in 2H or 4H, truck should not move.	a. Light does not come on before start up. Light re-mains on after parking brake is released. b. Parking brake inoper-able or un-able to hold truck.

TM 9-2320-289-10

Table 2-1. Preventive Maintenance Checks and Services for Model CUCV

Item No.	Interval	Location / Item to Check/ Service	Crewmember Procedure	Not Fully Mission Capable If:
12	Before	Brake System Continued	DRIVER NOTE Pedal should travel 1 to 1-1/2 inch before brakes take hold. After brakes take hold pedal may exceed the 1 to 1-1/2 inch travel. This is normal. c. Check brake pedal travel. With truck at idle, transfer in 2H, transmission in D, allow truck to move forward. As truck moves, slowly depress brake pedal.	c. Brakes will not stop truck.

2-16 Change 4

Table 2-1. Preventive Maintenance Checks and Services Model CUCV

Item No.	Interval	Location / Item to Check Service	Crewember / Procedure	Not Fully Mission Capable If:
			DRIVER	
13	During	Controls and Indicaters	Monitor all gages and warning lights during operation.	Warning light lights or gage drops below normal reading.
14	During	Brakes	Check brakes for pulling, grabbing.	Brakes pull or grab.
15	During	Steering	Be alert for excessive sway, leaning to one side, or unstable handling.	Handling is unstable.
16	During	Power Train	Be alert for unusual noises or vibrations from transmission, transfer, differentials, propeller shafts, axle shafts, or wheels.	Unusual noise or vibration detected.

TM 9-2320-289-10

Table 2-1. Preventive Maintenance Checks and Services for Model CUCV

Item No.	Interval	Location Item to Check Service	Crewmember Procedure	Not Fully Mission Capable If:
17	After	Under Hood	DRIVER NOTE Transmission fluid level should be checked with engine run-ning, parking brake set, trans-mission shift lever in "P" and ve-hicle on level ground, Fluid level should read between "ADD" and "FULL" on the dip-stick. Check transmission fluid level. If level is below "ADD", add suf-ficient fluid to bring the level be-tween "ADD" and "FULL" mark. NOTE After completing transmission fluid level check, turn off en-gine. If leakage if detected, further inves-tigation is needed to de-termine the location and cause of the leak.	

2-18 Change 4

Table 2-1. Preventive Maintenance Checks and Services Model CUCV

Item No.	Interval	Location / Item to Check/Service	Crewmember Procedure	Not Fully Mission Capable If:
18	After	Left Front, Sie Exterior	DRIVER a. Check underneath vehicle for evidence of fluid leakage.	a. Any brake fluid leakage, Class III engine, transmission, transfer, differential oil, coolant, fuel, or powersteering fluid leak.
			b. Visually check left side of vehicle for obvious damage that would impair operation.	b. Any damage that will prevent operation.
19	After	Left Side Tires	Visually check tires and wheels for inflation, cuts, gouges, bulges, serviceability and presence of lug nuts.	Tire unserviceable or one or more lug nuts missing.
20	After	Mirror	NOTE Vehicle operation with damaged or missing outside rear-view mirrors may violate AR 385-55 (M1010, M1028, M1028A1, M1028A2, and M1028A3 only). Check presence, cracks and serviceability of mirror operation.	

Change 5 2-19

TM 9-2320-289-10

Table 2-1. Preventive Maintenance Checks and Services for Model CUCV

Item No.	Interval	Location Item to Check Service	Crewmember Procedure	Not Fully Mission Capable If:
21	After	Rear Ex-terior	DRIVER NOTE If leakage is detected, further inves-tigation is needed to de-termine the location and cause of the leak. a. Check underneath vehi-cle for evidence of fluid leakage. b. Visually check rear of vehicle for obvious damage that would impair operation. (M1010 Only) c. Check operation of rear doors, han-dles and latching mechanisms. Check for loose or missing compo-nents. Doors should not bind and should close se-curely when latched shut.	a. Any brake fluid leak-age, Class III engine, transmis-sion, transfer, differen-tial oil, power steering fluid, coolant or fuel. b. Any damage that will prevent op-eration. c. Rear door han-dles and latching mechanisms d onot operate prop- erly or missing.

2-20 Change 4

Table 2-1. Preventive Maintenance Checks and Services Model CUCV

Item No.	Interval	Location / Item to Check/Service	Crewmember Procedure	Not Fully Mission Capable If:
			DRIVER	
2 2	After	Right Side Tires	Visually check tires and wheels for inflation, cuts, gouges, bulges, serviceability and presence of lug nuts.	Tire unserviceable or one or more lug nuts missing.
2 3	After	Spare Tire	Visually check spare for cuts, gouges, cracks, or bulges. Remove all penetrating objects.	Tire unserviceable or flat.
2 4	After	Mirror	NOTE Vehicle operation with damaged or missing outside rear-view mirrors may violate AR 385-55. Check presence, cracks and serviceability of mirror.	
2 5	After	Right Front, Side Exterior	NOTE If leakage is detected, further investigation is needed to determine the location and cause of the leak. a. Check underneath vehicle for evidence of fluid leakage. b. Visually check right side of vehicle for obvious damage that would impair operation.	a. Any brake fluid leakage, Class III engine, transmission, transfer, differential oil, coolant or fuel, Class II power steering fluid. b. b. Any damage that will prevent operation.

Change 4 2-21

TM 9-2320-289-10

Table 2-1. Preventive Maintenance Checks and Services for Model CUCV

Item No.	Interval	Location — Item to Check/ Service	Crewmember Procedure	Not Fully Mission Capable If:
			DRIVER	
26	After	Engine Oil Level	CAUTION Do not permit dirt, dust or grit to enter engine oil dip-stick tube. Internal engine damage will result if engine oil becomes contaminated. Do not overfill engine crank-case. Damage to engine will result. Check engine oil level. Level should be between "ADD" and "FULL'. If level is below "ADD", add oil to bring level between the ADD and FULL marks.	

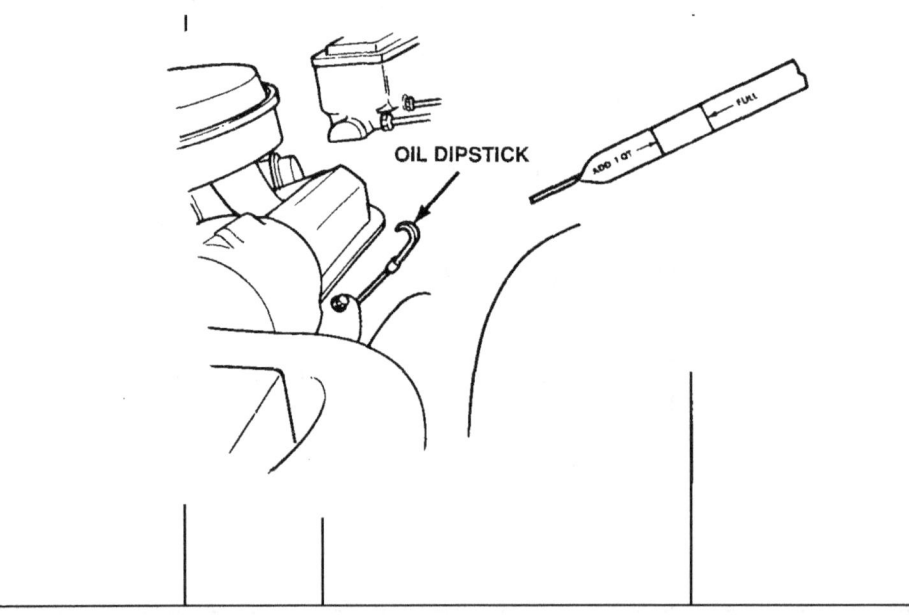

2-22 Change 4

TM 9-2320-289-10

Table 2-1. Preventive Maintenance Checks and Services Model CUCV

Item No.	Interval	Location / Item to Check/Service	<u>Crewmember</u> Procedure	Not Fully Mission Capable If:
27	After	Power Steering Fluid	<u>DRIVER</u> <u>CAUTION</u> **Do not permit dirt, dust, or grit to enter power steering reservoir. Damage to power steering system will result if power steering fluid becomes contaminated.** **Do not overfill power steering reservoir. Damage to power steering system will result.** Check fluid in power steering reservoir. Fluid should be between "HOT" and "COLD" marks. Add fluid if level is below "COLD" mark.	Class III leak.

Change 5 2-23

TM 9-2320-289-10

Table 2-1. Preventive Maintenance Checks and Services for Model CUCV

Item No.	Interval	Location Item to Check/Service	Crewmember Procedure	Not Fully Mission Capable If:
28	After	Power Steering Lines and Fittings	DRIVER **WARNING** **Notify organizational mechanic if power steering system has Class III leak. Loss of power could occur if this condition exists.** Visually check power steering lines and fittings for leaks or damage.	Any Class III leak.
		Master Cylinder	NOTE Ensure each reservoir is at proper level (see LO 9-2320-289-12). Visually check master cylinder, lines for leaks and security of cover. MASTER CYLINDER COVER	Any leak or cover missing.

2-24 Change 5

TM 9-2320-289-10

Table 2-1. Preventive Maintenance Checks and Services Model CUCV

Item No.	Interval	Location Item to Check/ Service	Crewmember Procedure	Not Fully Mission Capable If:
29	After	Cooling System	DRIVER **WARNING** **If engine has been recently operated on, do not remove radiator cap to check coolant level. Cooling system is under pressure and escaping steam or coolant can cause burns.** **Do not remove radiator cap to perform this task.** **CAUTION** **Overheating, caused by lack of coolant, will cause engine damage.** a. Check coolant level in see through coolant recovery tank. Level should be at or above the "FULL COLD" line. Add coolant if below the "FULL COLD" line. b. Inspect radiator hoses for leakage.	b. Class III leakage evident.

Change 42-25

TM 9-2320-289-10

Table 2-1. Preventive Maintenance Checks and Services for Model CUCV

Item No.	Interval	Location Item to Check/ Service	Crewmember Procedure	Not Fully Mission Capable If:
30	After	Lights	DRIVER NOTE Vehicle operation with dam-aged or inoperable headlightsmay violate AR 385-55. Check for presence and opera-tion of service drive, turn sig-nal, tail/stop light, blackoutmarker, blackout drive and sidemarker lights.	
31	After	Horns	DRIVER NOTE Operation of vehicles with in-opera-tive horn may violate AR 385-55. Check operation of horns if tac-tical situation permits.	
32	After	Wind-shield and Wipers	DRIVER NOTE Operation of vehicles withdamaged windshield may vio-late AR 385-55.a. Check windshield for dam-age that would impair opera-tor's vision. NOTE Vehicle operation with inopera-tive wipers may violate AR 385-55.b. Check windshield wiper andblade for presence and dam-age.	a. Windshield iscracked sufficiently toimpair opera-tor's vi-sion.

2-26 Change 4

TM 9-2320-289-10

Table 2-1. Preventive Maintenance Checks and Services Model CUCV

Item No.	Interval	Location / Item to Check Service	Crewmember Procedure	Not Fully Mission Capable If:
33	Weekly	Tires, Hubs and Wheels	**WARNING** Operating a vehicle with a tire in an under-inflated condition or with a questionable defect, may lead to prema-ture tire failure and may cause equipment damage, injury or death to personnel. a. Check tire tread depth and wear. Check for any cuts, cracks, gouges or bulges. Check for valve stem cap.	a. Tire tread depth is 1/8 inch (3.17 mm) or less or tread is worn to height of tread wear indicator. Any cut, gouge or crack that ex-tends to cor d body or any bulges.

WEAR INDICATORS

Change 4 2-27

TM 9-2320-289-10

Table 2-1. Preventive Maintenance Checks and Services for Model CUCV

Item No.	Interval	Location Item to Check/Service	Crewmember Procedure	Not Fully Mission Capable If:
33	Weekly	Tires, Hubs and Wheels Continued	b. Check locking hub for freedom of movement and for proper position (locked or unlocked) as required for either -two or four - wheel - drive operation.	

LUG NUTS

HUB KNOB

2-28 Change 4

TM 9-2320-289-10

Table 2-1. Preventive Maintenance Checks and Services Model CUCV

Item No.	Interval	Location Item to Check/ Service	Crewmember Procedure	Not Fully Mission Capable If:
3 3	weekly	Tires, Hubs and Wheels Continued	NOTE Maximum tire pressure for all trucks except M1009, must not exceed 80 psi or (552kPa). Maximum tire pressure for MI009 must not exceed 35 psi or (241 kPa). Gage tire for correct air pres-sure using the chart below. Ad-just pressures as necessary.	

Model	Front Rear		Spare
M1008	45 psi (310 kPa)	65 psi (448 kPa)	65 psi (448 kPa)
M1008A1	45 psi (310 kPa)	65 psi (448 kPa)	65 psi (448 kPa)
M1009	35 psi (241 kPa)	35 psi (241 kPa)	35 psi (276 kPa)
M1010	45 psi (310 kPa)	80 psi (552 kPa)*65 psi (448 kPa)*65 psi (448 kPa)	80 psi (552 kPa)
M1028	45 psi (310 kPa)		80 psi (552 kPa)
M1028A1	45 psi (310 kPa)		80 psi (552 kPa)
M1028A1	45 psi (310 kPa)	45 psi (310 kPa)	45 psi (310 kPa)
M1028A3	45 psi (310 kPa)	45 psi (310 kPa)	45 psi (310 kPa)
M1030	45 psi (310 kPa)	65 psi (448 kPa)	80 psi (552 kPa)

'Rear tire pressure with shelter loaded is 80 psi (552 kPa)

TM 9-2320-289-10

Table 2-1. Preventive Maintenance Checks and Services for Model CUCV

Item No.	Interval	Location Item to Check/ Service	Crewmember Procedure	Not Fully Mission Capable If:
3 4	Weekly	Doors and Windows	DRIVER CAUTION Do not open or close M1009 tailgate unless window is in full open (lowered) position. Opening or closing tailgate with window partly open may cause breakage of window. About 1/4 inch (.7 cm) of glass may remain exposed in full open position. a. Check operation of cab door and window. b. Inspect all doors and win-dews for indication of rust or cracks and/or breaks.	b. Any rusted through condition, cracks or breaks that would effect vehicle operation.
3 5	Weekly	Tailgate	Check tailgate for rust-through condition and/or damage and if tailgate latches securely.	Any rusted through condition or damage that effects vehicle operation.

2-30 Change 4

Table 2-1. Preventive Maintenance Checks and Services Model CUCV

Item No.	Interval	Location Item to Check/ Service	Crewrmember Procedure	Not Fully Mission Capable If:
			DRIVER	
36	Weekly	Exhaust System	Check exhaust system forcracks, rusted through condi-tion of pipes, muffler, or looseclamps or hangers.	Cracked or rustedthrough pipes or muf-fler.
37	Weekly	Shock Ab-sorbers	Visually inspect shock absorb-ers for leaks, damage and se-curity of mounting.	Class III leaks ordamage is evident.
38	Weekly	Cargo Cover	(M1008/M1008A1 w/cargo kitor M1008/M1008A1, M1009 w/win-terization kit)a. Check turn buttons andscrews securing canvas forse-curity of mounting.b. Check canvas compo-nents for rips and tears.	

Change 42-31

Table 2-1. Preventive Maintenance Checks and Services for Model CUCV

Item No.	Interval	Location Item to Check/Service	Crewmember Procedure	Not Fully Mission Capable If:
38	Weekly	Cargo Cover Continued	<u>DRIVER</u> (M1008/M1008A1 w/winterization kit) c. Check air inlet cover and filter for damage, dirt or obstruction. If dirty or clogged clean as necessary; if damaged notify organizational mechanic. AIR INLET COVER AIR EXHAUST VENTS d. Check turn buttons and screws securing radiator cover for security and mounting. e. Check air exhaust vent for freedom of operation and security of mounting. f. Check security and condition of plywood fasteners and components.	

Table 2-1. Preventive Maintenance Checks and Services Model CUCV

Item No.	Interval	Location Item to Check/Service	<u>Crewmember</u> Procedure	Not Fully Mission Capable If:
39	Weekly	Patient Compartment	<u>DRIVER</u> (M1010 ONLY) a. Check operation of cab passageway door. b. Check condition and operation of litter berths, litter wall brackets and tie-down straps. c. Check condition and operation of blackout curtains, hold-down clamps, and attendant's seat. d. Check condition and operation of domelight and focus lights. e. Check condition and operation of rear compartment doors, door windows, hold open devices and blackout switches.	b. One or more litter berths damaged or missing. c. One or more blackout curtains missing. d. Domelight and/or one or more focus lights missing or unserviceable. e. Any blackout switch inoperable.

Labels: DOMELIGHT, FOCUS LIGHTS, LITTER WALL BRACKETS, LITTER BERTHS, LITTER BERTHS TIE-DOWN STRAPS, HOLD OPEN DEVICES, HOLD OPEN DEVICES, HOLD-DOWN CLAMPS

TM 9-2320-289-10

Table 2-1. Preventive Maintenance Checks and Services for Model CUCV

Item No.	Interval	Location Item to Check/Service	Crewmember Procedure	Not Fully Mission Capable If:
40	Weekly	Air Conditioner	(M1010 only) Check air conditioner air inlet filter for obstruction and serviceability.	Air inlet blocked; subject to climatic or mission requirements.

2-34 Change 4

Table 2-1. Preventive Maintenance Checks and Services Model CUCV

Item No.	Interval	Location / Item to Check/ Service	Crewmember / Procedure	Not Fully Mission Capable If:
41	Weekly	Patient Compartment Heater	<u>DRIVER</u> (M1010 Ambulance) **NOTE** • Vehicle must be running to perform the following checks. • Heater is located behind removable panel on right side of vehicle. Perform the following if climatic conditions or mission require personnel heater. **NOTE** Before cleaning personnel heater air inlet filter screen, be sure truck is parked on level ground or nose of truck is lower than the rear. Engine and heater must be turned off. Do not use high volume hoses.	

TM 9-2320-289-10

Table 2-1. Preventive Maintenance Checks and Services for Model CUCV

Item No.	Interval	Location Item to Check/ Service	Crewmember Procedure	Not Fully Mission Capable If:
41	Weekly	Patient Compartment Heater Continued	DRIVER a. Inspect heater air inlet for dirt, damage, or obstruction. If filter is dirty clean as necessary. **NOTE** Visible smoke during the first three minutes of operation is normal. Blower operation for up to three minutes after heater is shut off is normal. **REMOVABLE PANEL** **HEATER AIR INLET** b. Check heater switches and controls for proper operation. Listen for unusual noises during operation. c. Check for exhaust leakage.	a. Air inlet damaged and heater is required. b. Mission requires heater and heater is inoperative. c. Mission requires heater and exhaust leakage is evident.

2-3 6 Change 4

Table 2-1. Preventive Maintenance Checks and Services Model CUCV

Item No.	Interval	Location Item to Check/Service	<u>Crewmember</u> Procedure	Not Fully Mission Capable If:
42	Weekly	Gas Particulate Filter Unit	<u>DRIVER</u> Check gas particulate filter unit for damage. Check for steady flow of air at hose outlets. Check filter unit heater control knobs for proper operation. When knobs are turned clockwise, a green light will come on indicating heater is working. If light does not come on, notify your supervisor or organizational mechanic.	

Change 42-3 7

TM 9-2320-289-10

Table 2-1. Preventive Maintenance Checks and Services for Model CUCV

Item No.	Interval	Location Item to Check/ Service	Crewmember Procedure	Not Fully Mission Capable If:
43	Weekly	Front Tow Hooks	Check presence and condition of tow hooks.	
44	Weekly	Air Cleaner	**WARNING** **If NBC exposure is suspected, all air filter media should be handled by personnel wearing protective equipment. Consult your unit NBC Officer or NBC NCO for appropriate handling or disposal instructions.** a. Visually check air cleaner cover and air cleaner assembly for security of mounting and damage. b. Visually check air cleaner polywrap for tears or dirt. If dirty, clean with hot soapy water and squeeze out excess water, then dip in light engine oil and squeeze out excess oil. c. Visually check air cleaner paper element for dirt, oil or damage.	a. Evidence of damage to air cleaner cap, body, or mounting that will allow unfiltered air to enter engine.

2-38 Change 4

TM 9-2320-289-10

Table 2-1. Preventive Maintenance Checks and Services Model CUCV

Item No.	Interval	Location / Item to Check/ Service	Crewmember Procedure	Not Fully Mission Capable If:
45	Weekly	Alternator Brackets	Visually check alternator brackets for cracks, damage or loose bolts.	Damaged or cracked brackets or loose bolts.
46	Weekly	Cooling System	a. Check fan and all pulleys for damage.	a. Fan blade or any pulley bent, broken, cracked, or loose.

Change 42-39

TM 9-2320-289-10

Table 2-1. Preventive Maintenance Checks and Services for Model CUCV

Item No.	Interval	Location Item to Check/Service	Crewmember Procedure	Not Fully Mission Capable If:
46	Weekly	Cooling System Continued	b. Check for loose, missing, broken, cracked, or frayed drivebelts. c. Check radiator for leaks, clogged or damaged fins.	b. Any drivebelt is loose, missing, broken, cracks to the belt fiber, has more than one crack (1/8 inch in depth or 50% of belt thickness) or has frays more than 2 inches long. c. Class III leak evident.

BELTS — All Except M1010

BELTS — M1010

2-40 Change 4

Table 2-1. Preventive Maintenance Checks and Services Model CUCV

Item No.	Interval	Location Item to Check/Service	Crewmember Procedure	Not Fully Mission Capable If:
47	Weekly	Heater	(Winterization Kit) a. Check warm air heater for security of mounting. b. Check fuel lines, filter and connections for cracks, breaks, or leaks. c. Check electrical cables and connections for security, frayed or broken wires.	 b. Class III fuel leak, cracks, broken lines or connections. c. Frayed or broken wires.

WARM AIR HEATER

TM 9-2320-289-10

Table 2-1. Preventive Maintenance Checks and Services for Model CUCV

Item No.	Interval	Location Item to Check/ Service	Crewmember Procedure	Not Fully Mission Capable If:
48	Weekly	Engine Coolant Heater	a. Check engine coolant heater for security of mounting.	
			b. Check fuel lines and connections for cracks, breaks, or leaks.	b. Class III fuel leaks, cracks, broken lines or connections.
			c. Check electrical cables and connections for security, frayed or broken wires.	c. Frayed or broken wires.
			d. Check coolant hoses for cracks, brakes or leaks.	d. Class III coolant leak.
			e. Check heater fuel pump and fuel filter for leaks or damage.	e. Class III fuel leak.

2-42 Change 4

TM 9-2320-289-10

Table 2-1. Preventive Maintenance Checks and Services Model CUCV

Item No.	Interval	Location Item to Check/ Service	Crewmember Procedure	Not Fully Mission Capable If:
49	Weekly	Heater Control Box	a. Check heater control box for security of mounting and electrical connections. b. Check operation of switches, lights and battery/interior lever.	 b. Switches inoperable and climatic conditions require heaters.
			[Diagram of Heater Control Box showing LIGHTS, BATTERY/INTERIOR LEVER, HEATER CONTROL BOX, and ELECTRICAL CONNECTIONS]	
50	Weekly	Cargo Compartment Dome Light	a. Check domelight for condition and operation in both normal and blackout modes. b. Check cables for frays or broken wires.	 b. Frayed or broken wires.

Change 42-4 3

TM 9-2320-289-10

Table 2-1. Preventive Maintenance Checks and Services for Model CUCV

Item No.	Interval	Location / Item to Check/Service	Crewmember Procedure	Not Fully Mission Capable If:
51	Weekly	Cargo Heater	a. Check heater for security of mounting. b. Check fuel lines and fittings for leaks, cracks or breaks. c. Check electrical cables and connections for security, frayed or broken wires.	b. Class III fuel leak, cracks, broken lines or connections. c. Frayed or broken wires.

2-4 4Change 4

Table 2-1. Preventive Maintenance Checks and Services Model CUCV

Item No.	Interval	Location — Item to Check Service	Crewmember Procedure	Not Fully Mission Capable If:
52	Weekly	Tow Pintle, Hooks, Electrical Connector and Step	a. Visually check pintle hook for looseness, dam-aged locking mech-anism, and presence of cotter pin (all except M1010 and M1031). b. Check presence and condition of tow hooks (all except M1010 and M1031). c. Check trailer electrical connector for damage (all except M1010 and M1031). d. Check condition and op-eration of rear step (1010 only).	

TM 9-2320-289-10

Table 2-1. Preventive Maintenance Checks and Services for Model CUCV

Item No.	Interval	Location Item to Check/ Service	Crewmember Procedure	Not Fully Mission Capable If:
53	Weekly	Battery	**WARNING** **Remove all jewelry such as rings, dog tags, bracelets, etc. If jewelry contacts battery terminal, direct short will result causing instant heating of tools, severe injury to personnel or damage to equipment.** a. Visually check batteries for defects such as burnt or corroded terminals, cracked or damaged case.	a. One or more batteries missing or unserviceable. Any terminal or cable loose or corroded. Any battery holddown is not secure.

2-46 Change 4

TM 9-2320-289-10

Table 2-1. Preventive Maintenance Checks and Services Model CUCV

Item No.	Interval	Location Item to Check/ Service	Crewmember Procedure	Not Fully Mission Capable If:
53	Weekly	Battery Continued	**NOTE** There is a charge indicator on top of manufacturer supplied Delco batteries. When indicator shows a green dot, battery is OK. If indicator shows a yellow or black dot, notify organizational maintenance. b. Check charge indicator of battery (manufacturer supplied battery). If indicator shows either yellow or black, notify organizational mechanic. **CHARGE INDICATOR** DARKENED INDICATOR (WITH GREEN DOT) MAY BE SLAVE STARTED DARKENED INDICATOR (NO GREEN DOT) MAY BE SLAVE STARTED LIGHT YELLOW OR BRIGHT INDICATOR DO NOT SLAVE START c. Check electrolyte level (military standard battery). If low add distilled water. If fluid is boiling notify organizational maintenance.	b. Indicator shows either yellow or black dot. One or more batteries missing or unserviceable. c. Fluid is boiling or one or more batteries missing or unserviceable.

Change 4 2-47

TM 9-2320-289-10

Table 2-1. Preventive Maintenance Checks and Services for Model CUCV

Item No.	Interval	Location Item to Check/ Service	Crewmember Procedure	Not Fully Mission Capable If:
54	Weekly	Fuel Filter	Check fuel filter for leaks or damage.	Class III leak evident.
55	Weekly	Slave Receptacle	Check slave receptacle for damage to receptacle or cover.	

FUEL FILTER

FRONT

2-48 **Change 4**

TM 9-2320-289-10

Table 2-1. Preventive Maintenance Checks and Services Model CUCV

Item No.	Interval	Location Item to Check/ Service	Crewmember Procedure	Not Fully Mission Capable If:
56	Monthly	Corrosion	Visually inspect for indication of corrosion or cracks and/or breaks.	Any corroded through condition or cracks or breaks that would effect vehicle operation.
57	Monthly	Tailgate	Check tailgate for corroded through condition and/or damage. If tailgate does not latch securely or is damaged, notify organizational maintenance.	Any corroded through condition or damage that would effect vehicle operation.
58	Monthly	Med Cross Plate	a. Check cross marking latches and hinge for preparation, security of mounting, damage or missing components. b. Inspect stowage compartment door hinge, seal and latch for proper operation, damage or missing components.	
59	Monthly	Shelter Mount Kit	Inspect shelter mounting bracketry for security of mounting and loose or missing bolts.	Any mounting bolt missing.
60	Monthly	Troop Seats	Troop Seats Kit a. Inspect troop seats for missing or damaged lockpins. b. Inspect troop seats and backrests for security of mounting.	b. Mission requires troop seats are inoperative or damaged.

☆ U.S. GOVERNMENT PRINTING OFFICE: 1992 — 643-025/40265   **Change 4**   2-49/2-50 blank)

Note: Pages 2-51 and 2-52 have been deleted.

Table 2-1.

Table 2-1. Operator/Crew Preventive Maintenance Checks and Services - Continued

B - BEFORE   D - DURING   A - AFTER   W - WEEKLY   M - MONTHLY

ITEM NO.	INTERVAL					ITEM TO BE INSPECTED PROCEDURE: CHECK FOR AND HAVE REPAIRED, FILLED OR ADJUSTED AS NEEDED	EQUIPMENT IS NOT READY/ AVAILABLE IF:
	B	D	A	W	M		
					●	k. Check all turnbuttons and screws securing canvas and radiator cover for security of mounting.	
					●	l. Check canvas components for rips or tears.	
					●	m. Check plywood components for security of screws and bolts.	
					●	n. Check plywood components for damage or broken pieces.	

Table 2-1. Operator/Crew Preventive Maintenance Checks and Services - Continued

Table 2-1. B - BEFORE D - DURING A - AFTER W - WEEKLY M - MONTHLY

ITEM NO.	INTERVAL					ITEM TO BE INSPECTED PROCEDURE: CHECK FOR AND HAVE REPAIRED, FILLED OR ADJUSTED AS NEEDED	EQUIPMENT IS NOT READY/ AVAILABLE IF:
	B	D	A	W	M		
					•	o. Check air inlet (16) and air exhaust vents (17) (M1008 and M1008A1 only) for freedom of operation and security of mounting. Check filters for dirt, damage, or obstruction. If dirty or clogged, clean as necessary. If damaged, notify your supervisor.	

TM 9-2320-289-10

Section III. OPERATE UNDER USUAL CONDITIONS

2-8.

2-8. START ENGINE

a. Adjust seats.

b. Adjust all mirrors.

c. Buckle seat belts.

d. Set parking brake (if not already set).

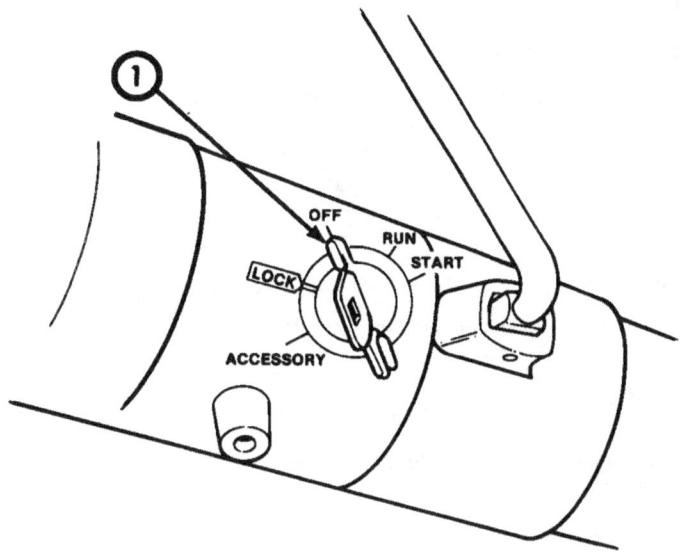

e. Insert key into starter ignition switch (1) and turn key to "RUN."

NOTE

DO NOT turn key to "START" position until WAIT light goes out.

f. When WAIT light goes out

(1) If temperature is more than 32°F (0°C), press accelerator pedal halfway to floor and hold.

(2) If temperature is less than 32°F (0°C), press accelerator pedal to floor and hold.

NOTE

If engine doaa not start after cranking shout 10 to 15 seconds, release ignition key. Wait five seconds, then attempt to atart engine again. If engine fails to atart, refer to Troubleshooting, Table 3-1 and notify your supervisor. DO NOT tow or push start truck.

g. Turn key in ignition switch to "START." When engine starts, release key and allow it to return to "RUN" position.

h. Move service lights/blackout toggle switch to "ON" position.

TA466606

Change 5   2-53

TM 9-2320-289-10

2-9. DRIVE TRUCK FORWARD

a. Start the engine. (See paragraph 2-8)

b. Check your instruments.

   (1) Voltmeter should show slight charge

   (2) Oil pressure, engine coolant temperature, WAIT, low coolant, and generator lights should be Out.

c. Shift transfer case control lever to appropriate range ("2H" for normal driving conditions).

d. Shift transmission gearshift lever to desired forward gear ("D" (Drive) for most normal driving conditions).

e. Release parking brake

   (1) Depress service brake pedal

   (2) Release parking brake by pulling on the T-handle brake release

f. Release service brake pedal.

g. Depress accelerator pedal gradually to move the truck.

2-10. STOP TRUCK

a Release accelerator pedal.

b. Depress and hold service brake pedal to bring truck to a complete stop.

2-11. DRIVE TRUCK IN REVERSE

NOTE

The truck's speedometer does not indicate reverse speeds. You must estimate your speed and use good judgement when driving in reverse.

a. Bring truck to a complete stop.

b. Keeping your foot on service brake pedal, shift transmission gearshift lever to "R" (Reverse),

c. Release service brake pedal and gradually depress accelerator pedal.

2-12 DRIVE TRUCK IN FOUR-WHEEL DRIVE

a. Shift from Two-wheel Drive to Four-wheel Drive.

CAUTION

Both locking hubs must be in same position before completing transfer case shift or damage to drive train components can occur.

   (1) Lock Locking Hubs.

      (a) Stop truck and set parking brake,

      (b) Shift transmission gearshift lever to "P" (Park).

TM 9-2320-289-10

NOTE

If hubs do not turn freely, drive the truck backward or forward a few feet and try again.

    (c) Turn each locking hub knob (1) clockwise until it stops and pointer (2) on the hub dial is lined up with the word "LOCK,"

(2) Shift Transfer Case from "2H" to "4H."

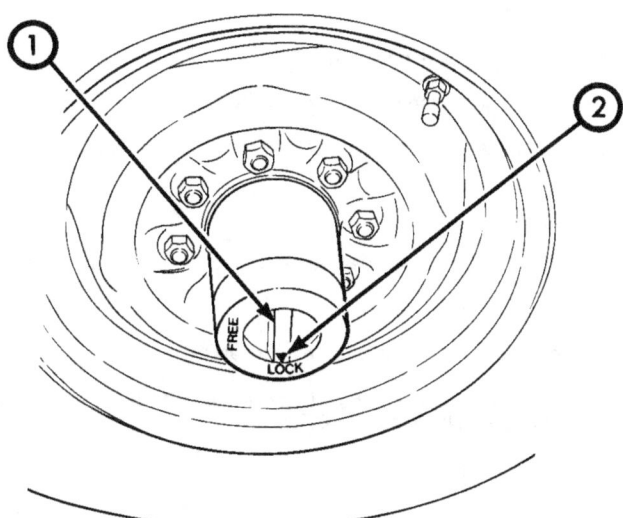

    (a) Shift transmission gearshift lever to "D" (Drive) or appropriate forward gear (reverse may be selected if absolutely necessary).

    (b) Release parking brake.

    (c) Drive truck forward, within speeds specified in paragraph 1-13.

    (d) Momentarily release accelerator, then firmly but not forcefully, shift transfer case control lever to "4H," according to the pattern indicated on transfer case control lever.

NOTE

If it is not possible to move truck to shift transfer case, it can be shifted by shifting transmission gearshift lever to "N" (Neutral) then shifting transfer case control lever firmly but not forcefully. If shifting the transfer case is difficult, it may be necessary to temporarily shift the transmission gearshift lever to "D" (Drive) then back to "N" (Neutral) and try again to shift transfer case.

(3) Shift Transfer Case from "2H" or "4H" to "4L."

    (a) Stop truck and set parking brake.

    (b) Shift transmission gearshift lever to "N" (Neutral),

    (c) Release parking brake.

    (d) Shift transfer case control lever firmly, but not forcefully, to "4L."

TA466607

TM 9-2320-289-10

    b. Return to Two-wheel Drive from Four-wheel Drive.

        (1) Return to "2H" from "4H."

            (a) Shift transmission gearshift lever to "D" (Drive) or appropriate forward gear

            (b) With truck in motion, momentarily release accelerator and, firmly but not forcefully, shift transfer case control lever according to pattern indicated on lever.

        (2) Return to "2H" or "4H" from "4L."

            (a) Stop truck and set parking brake

            (b) Shift transmission gearshift lever to "N" (Neutral)

            (c) Shift transfer case control lever to "2H'" or "4H" according to pattern indicated on lever.

        (3) Unlock Locking Hubs.

            (a) Ensure that transfer case has been returned to the "2H" position

            (b) Set parking brake

            (c) Shift transmission gearshift lever to "P" (Park)

            (d) Turn each locking hub knob counterclockwise until it stops and pointer on dial is lined up with the word "FREE."

2-13. PARK TRUCK AND SHUT DOWN ENGINE

NOTE

DO NOT shift transmission gearshift lever to "P" (Park) on a hill before setting parking brake. This puts force on the transmission and makes it difficult to shift the transmission gearshift lever out of "P" (Park). Make sure the transfer case is in gear.

    a. Bring truck to a complete stop.

    b. Set parking brake by depressing service brake pedal with your right foot, and holding it while setting parking brake with your left foot.

WARNING

If transfer case control lever is in "N" (Neutral), transfer case is disengaged and shifting the transmission gearshift lever to "P" (Park) WILL NOT stop the truck from moving.

    Ensure that transfer case is in gear.

    Shift transmission gearshift lever to "P" (Park).

    Move all switches to the "OFF" position.
c.
    Turn key in ignition to "LOCK" and remove key.
d.

e.

f.

TM 9-2320-289-10

2-13.1 TOWING WITH M1009

WARNING

If the tailgate and/or tailgate window is required to be opened while moving or towing, extreme caution is required. Exhauat fumes may enter resulting in injury or death to personnel.

NOTE

Although not recommended, if your M1009 requires the tailgate and/or tailgate window to be open while moving or towing, the following procedure is required.

a. Close the door windows.

NOTE

Use either the heating or cooling system as environment permits.

b. Set the fan to intermediate or high speed (see para 2-20c).

c Set upper control lever to any position except "OFF' (see para 2-20a).

d. Open both outside air vents fully (see para 2-22).

PIN: 053018-001

2-14. TOW A TRAILER/AIRCRAFT

a. Towing Capacities of CUCV Models.

(1) M1009. Equipped to tow M416 Series 1/4 ton trailers with maximum towed load of 1200 pounds for both cross-country and highway travel. Load on tongue should not exceed 100 pounds.

(2) FA1008, M1008A1, M1028, M1028A1, M1028A2, and M1028A3, Equipped to tow M101 Series 3/4 ton trailers with maximum towed load of 3100 lbs. Load on tongue should not exceed 300 lbs.

b. Guidelines for Towing Aircraft Which Exceed Maximum Towed Load Weight.

(1) Speed will not exceed 5 mph.

(2) Truck will be operated with transmission in low range (" 1") and transfer case in "2H."

(3) Gross load will not exceed 10,000 pounds for M1009, 15,000 pounds for M1008, M1008A1, M1028, M1028A1, M1028A2, and M1028A3.

(4) Tongue load will not exceed 100 pounds for M 1009, and 300 pounds for M1008, M1008A1, M1028, M1028A1, M1028A2, and M1028A3.

(5) Operation will be on a level, hard surface.

c. Guidelines for Towing a Trailer on Hills.

(1) When going up or down long, steep hills, downshift transmission to lower gear range and reduce truck speed to 45 mph (72 kph) or below.

(2) if you must park truck with loaded trailer on a hill, follow these steps:

(a) Apply service brakes.

(b) Have someone place chocks (1) under downhill side of trailer wheels.

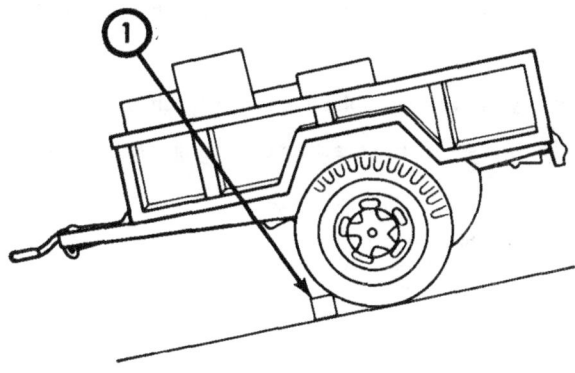

(c) When trailer wheel chocks (1) are in place, slowly release service brakes until chocks take up the load of trailer.

TM 9-2320-289-10

(d) Set parking brake.

<u>CAUTION</u>

DO NOT shift transmission gearshift lever to "P" (Park) until the trailer wheels are chocked and the parking brake is set or it may ba hard to shift out of "P" (Park).

(e) Shift transmission gearshift lever to "P" (Park).

(3) When starting from a parked position on a steep hill, follow these steps:

(a) Start engine in "P" (Park).

(b) Apply service brakes and hold down brake pedal.

(c) Shift transmission gearshift lever into gear and release parking brake

(d) Drive truck slightly uphill and remove chocks before driving away.

2-15. OPERATE LIGHTS

a. Service Lights.

NOTE

Headlights, horn, turn signals, and hazard warning flasher will not operate unless service lights/blackout toggle switch is set to "SERVICE LIGHTS ON."

(1) On. Pull out and move service lights/blackout toggle switch to "SERVICE LIGHTS ON," then pull out headlight switch to turn on headlights as necessary.

(2) Off. Pull out and move service lights/blackout toggle switch to "ALL OFF," then push in on headlight switch.

b. Blackout Drive Lights.

(1) On. Pull out and move service lights/blackout toggle switch to "BLACKOUTON,"then pull out and push up, and release blackout drive light toggle switch to "BLACKOUT DRIVE LIGHT ON." Left toggle switch will return to center position when released.

(2) Off. Pull out and move service lights/blackout toggle switch to "ALL OFF."

c. Blackout Markers.

(1) On. Pull out and move service lights/blackout toggle switch to "BLACKOUT ON."

(2) Off. Pull out and move service lights/blackout toggle switch to "ALL OFF." 2-16. OPERATE

HEADLIGHT DIMMER SWITCH

a. Pull out and move service lights/blackout toggle switch to "SERVICE LIGHTS ON" and turn on headlights.

2-58

TM 9-2320-289-10

b. Turn on high beams.

(1) With headlights on, pull turn signal control lever (1) toward you.

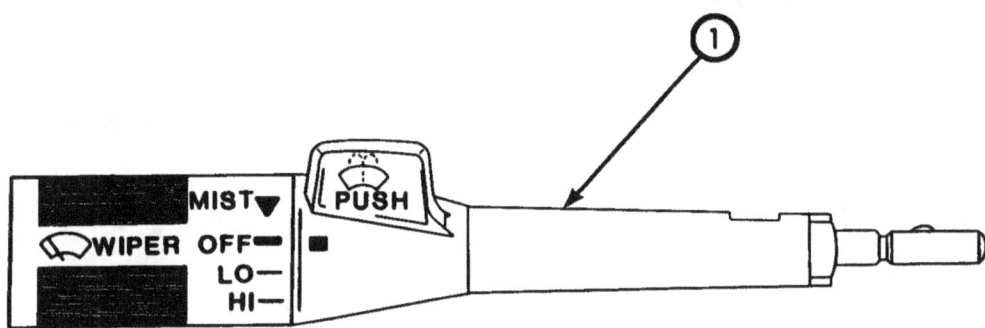

(2) Ensure that high beam indicator comes on.

c. Return to low beams.

(1) Pull turn signal control lever toward you.

(2) Ensure that high beam indicator goes out.

2-17. OPERATE TURN SIGNAL INDICATOR

NOTE

- Turn signal indicator will not oparate unless service lights/blackout toggle switch is set to "SERVICE LIGHTS ON."
- Turn signal indicator will automatically turn itself off once turn ia completed.

s. To indicate a left-hand turn. oush turn signal control lever down. Ensure that left turn indicator on dashboard is flashing.

b. To indicate a right-hand turn, push turn signal control lever up. Ensure that right turn indicator on dashboard is flashing.

TM 9-2320-289-10

2-18. OPERATE WINDSHIELD WIPER/WASHER

NOTE

Windshield wipers ara electric and will not operate if ignition switch is off.

    a. To turn wipers on, twist knob (1) on end of turn signal control lever (3) toward front of truck to "LO" (4) for normal wiper speed or"HI" (5) for a more rapid wiper speed.

    b. To turn off wipers, twist knob (1) toward rear of truck to "OFF" (6).

    c. If a light mist requires only one wipe by wipers, this can be done by twisting knob (1)toward the rear of the truck to "MIST" (7).

    d. To operate windshield washer, push thumb plate (2) toward front of truck. Windshield wipers will automatically come on.

2-19. OPERATE HORN

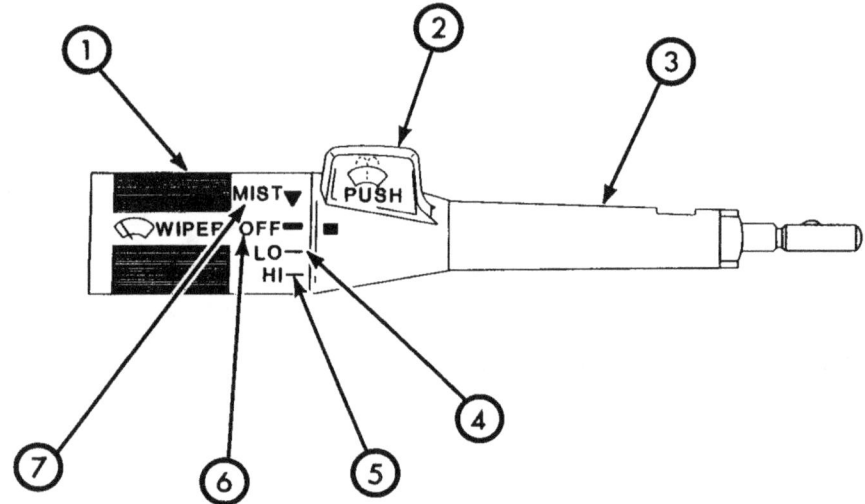

NOTE

Horn will not operate unless service lights/blackout toggle switch is set to "SERVICE LIGHTS ON."

Depress centerpiece (1) of steering wheel to sound horn.

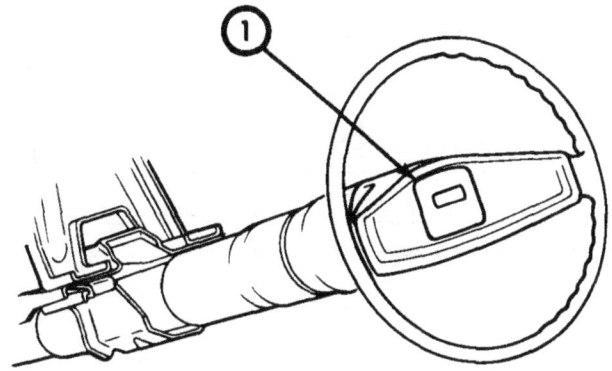

2-20. OPERATE HEATER/ DEFROSTER

NOTE

Always keep air path under seat clear of objects.

a. Operate Uppar Lever (2).

(19 With lever (2) in "OFF" (1) position, unheated air comes out of heater outlet.

(2) With lever in "HEATER" (3) position, heated air flows from heater outlet.

(3) With lever in "'DEF'" (4) position, heated air flows to defroster outlets.

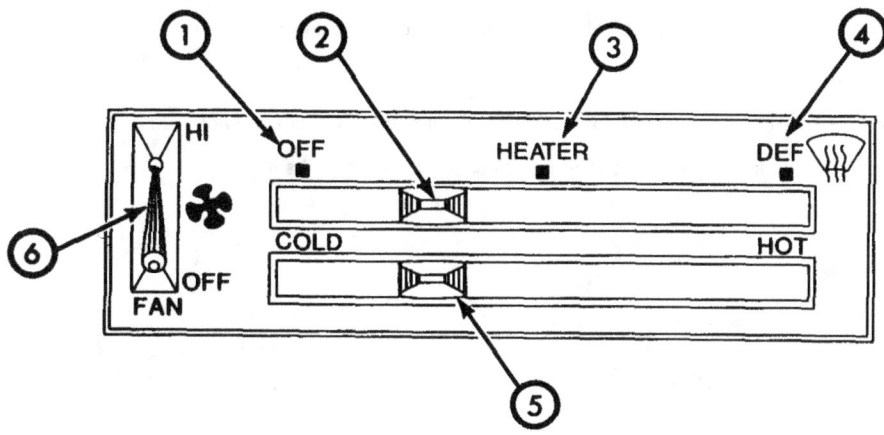

b. Operate Lower Lever (5).

(1) Move lever (5) from left to right to regulate heat.

(2) Full right position provides the most heat.

c. Operate Fan (6).

(1) Move fan lever (6) from "OFF" through notches to "HI" position to turn on fan motor and regulate force of air.

(2) Move fan lever (6) to "OFF" position to turn off fan motor.

TM 9-2320-289-10

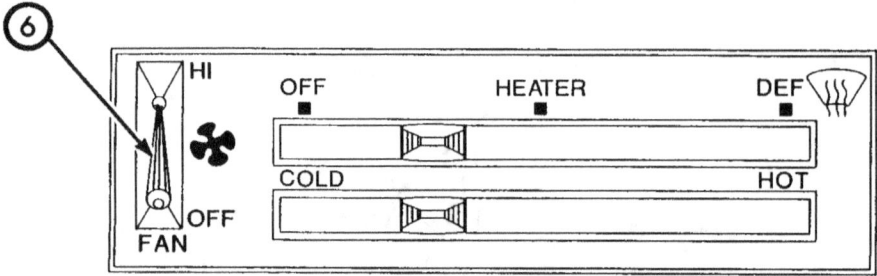

d. Operate Heatar/Dafrostar Under Winter Snow and Icy Conditions.

(1) Clear snow and ice from hood and air inlet in front of windshield.

(2) Run fan (6) on "HI" for a few moments before driving off to clear intake ducts of snow and moisture.

2-21. OPERATE HAZARO WARNING FLASHER

NOTE

Hazard warning lights will not work unless service lights/blackout toggle switch is set to "SERVICE LIGHTS ON."

a. To turn hazard warning flashers on, push in on button located on steering column just behind steering wheel.

b. To turn hazard warning flashers off, pull out button on steering column

2-22. OPERATE AIR VENTS

a. Knobs (1) to operate air vents are located under the instrument panel on each side of cab.

b. To open vents, pull knobs (1) toward you; to close vents, push knobs in.

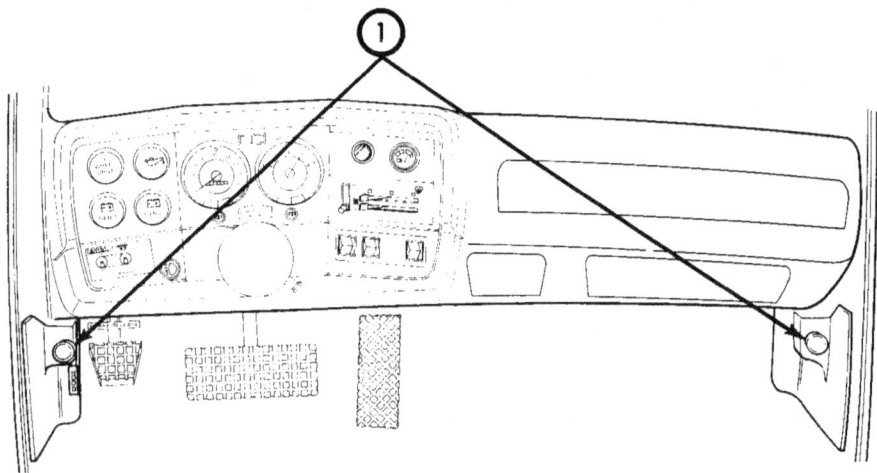

TM 9-2320-289-10

2-23. OPERATION TAILGATE (M1009)

CAUTION

- Tailgate or glass will window must be completely lowered before tailgate can be opened or window glass may break. When window glass is completely lowered, ap-proximately 1/4 Inch (.7 cm) of remain exposed.
- Glass does not seat completely in door. DO NOT drop heavy objects on glass or glass may break.

a. Lower Tailgate Window.

Unlock tailgate using ignition key.

PuN out window (1) regulator handle (1) at end indicated by the arrows.

Turn handle (2) (1) counterclockwise until desired position is reached, then rotate handle clockwise to a horizontal (3) position and snap into place.

CAUTION

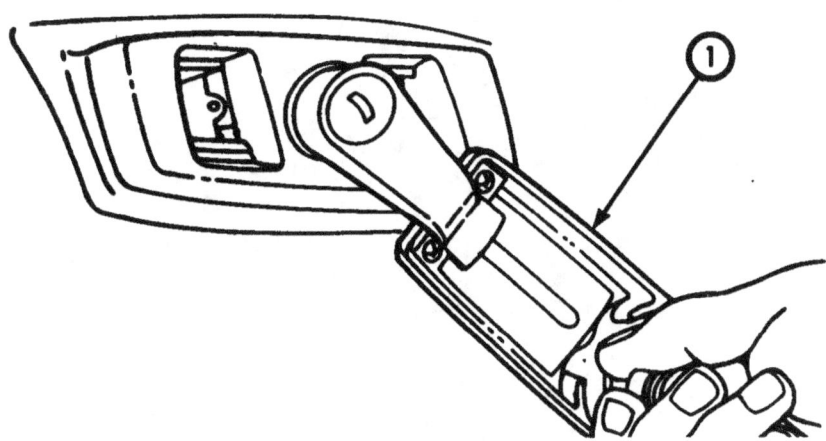

DO NOT attempt to toad large or heavy items into open window of truck. Window does all the way down into tailgate, so part of glass ia exposed end mey break

b. Open Tailgate.

(1) Completely lower window, as described above.

(2) Lift release handle located on the inside of tailgate just below window.

(3) Pull tailgate open.

c. Close Tailgate. into position and close firmly.

d. Rasie Window.

PuN out window regulator handle (1) at end indicated by arrows.

Tum handle (1) clockwise to raise window.

(1) When desired window height is reached, lock tailgate with key, rotate handle (1) counterclock-
(2) wise) and snap into place.
TA466613
(3)

Change 5 2-63

TM 9-2320-289-10

2-24. OPERATE WEAPONS MOUNT (ALL EXCEPT M1010)

NOTE

Single mounts are located behind seat on all trucks except Ml009. M1009 has a double mount behind driver's seat.

a. Move latch (1) at top of mount to either side.

b. Place weapon stock in boot (2) on floor and push barrel of weapon into top portion

c. Move latch (1) to hold barrel of weapon in place.

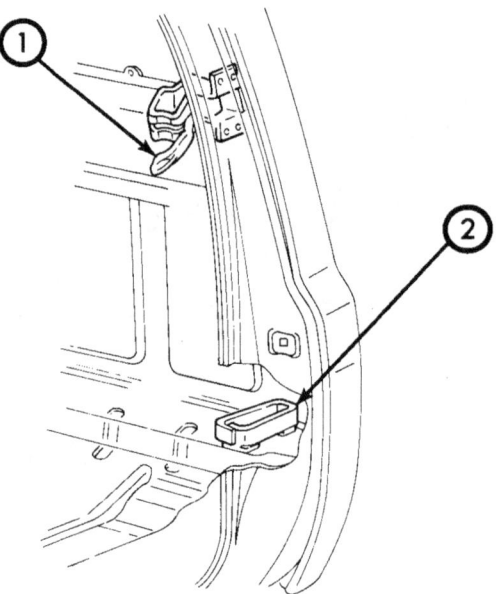

All Except M1009 and M1010

2-64

TM 9-2320-289-10

Section IV. OPERATE AMBULANCE   PECULIAR COMPONENTS

2-25. OPERATE GAS-PARTICULATE FILTER SYSTEM (M1010)

NOTE

- There ere seven hookup positions for the gas-particulate filter system, six of which may be used at the same time. Five ara in the patient compartment, two in the cab.

- The GPFU is designed to operate with the M25 Series protective mask.

a. Put on protective mask, clear, and seal it.

b. Put on helmet.

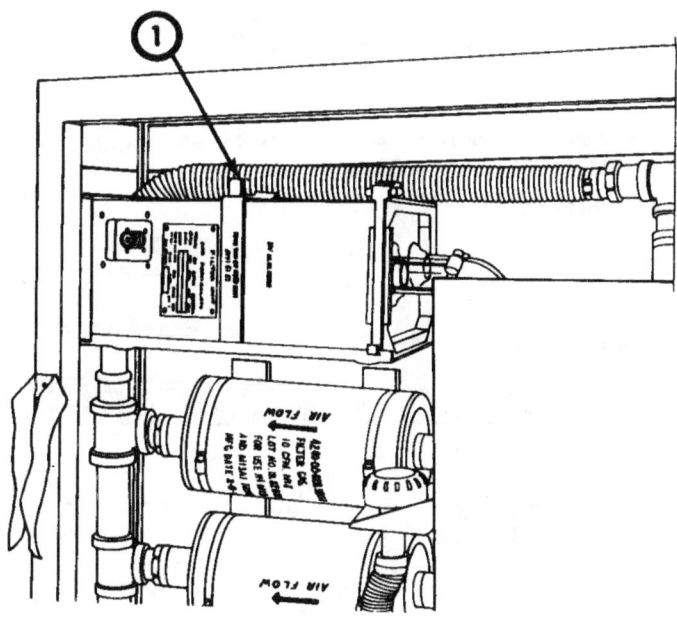

NOTE

Spring clip (1) must be removed for air filter assembly to function.

c. Remove spring clip (1) from intake opening of gas-particulate filter unit.

TA466615

2-65

TM 9-2320-289-10

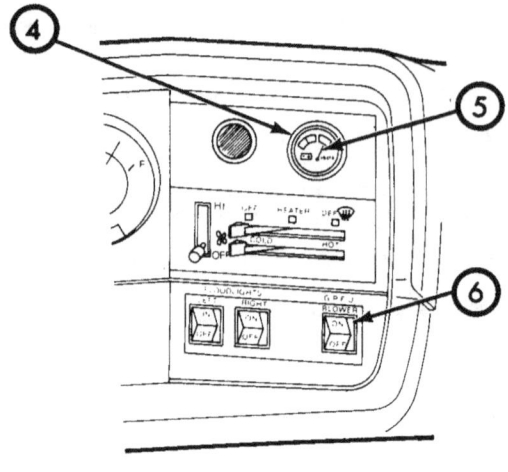

CAUTION

Frequently observe voltmeter (4) during GPFU operation. if needle (5) falls into yellow zone, start angine and recharge battaries.

d. Turn GPFU switch (6) to "ON."

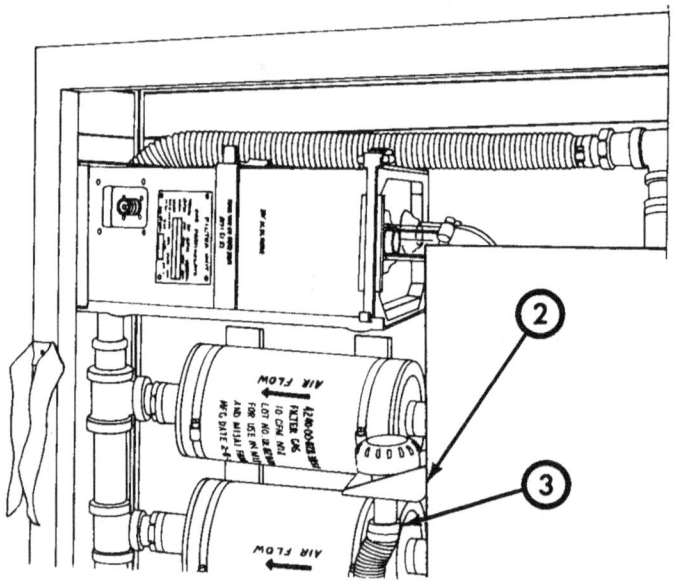

WARNING

Under extreme cold conditions, danger of frostbite exists. Put on protective mask, but DO NOT connect air duct to mask until heater has been on for 15 minutes.

e. Remove air duct hose breakaway socket (3) from mounting (2),

f. Connect air duct hose breakaway socket (3) to protective mask cannister.

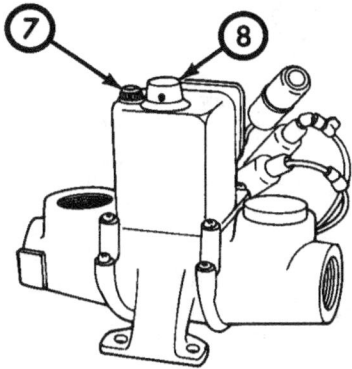

g. If air is too cold, turn M3 heater control knob (8) clockwise until indicator light (7) comes on.

h. Turn M13 heater control knob (8) counterclockwise for more normal air temperature.

i. When M3 heater is no longer required, turn heater control knob (8) counterclockwise until it stops, to turn off heater.

j. When no longer required, remove protective mask and stow air duct hose breakaway socket (3).

k. Push GPFU switch (4) to "OFF."

2-26. OPERATE AIR CONDITIONER (M1010)

NOTE

o For maximum efficiency, close all windows and vents after the first few minutes of o

e "OUTSIDE-A/C NORMAL OR VENT" setting passes outside air through air conditioning outlet. It is the position most used.

l "INSIDE-A/C MAX setting recirculate cool air from inside the truck.

a. Move air selection lever (3) from "CLOSED-OFF" to either "INSIDE-A/C MAX" or "OUTSIDE-A/C NORMAL OR VENT."

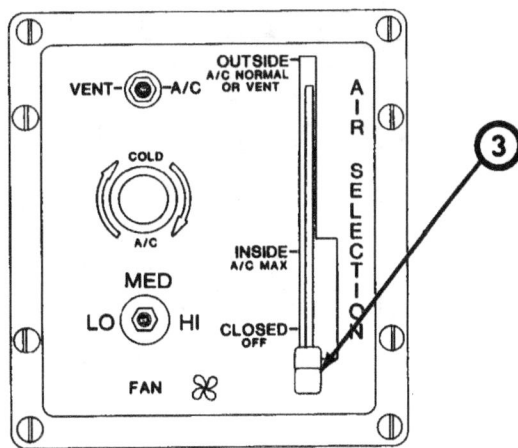

TM 9-2320-289-10

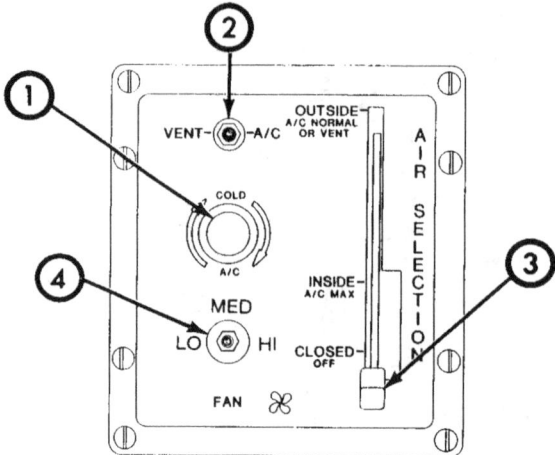

b. Move "VENT-A/C" toggle switch (2) to "VENT" for circulation of outside air or "A/C" for circulation of cooled air.

c. Move fan switch (4) to any desired setting from "LO" to "HI," as required.

d. Move "A/C-COLD" knob (1) clockwise for colder temperature setting,

e. Move "A/C-COLD" knob (1) counterclockwise for warmer temperature setting

f. When air conditioner operation is no longer required, move air selection lever (3) to "CLOSED-OFF."

g. Move fan switch (4) to "LO."

2-27. OPERATE PATIENT COMPARTMENT HEATER (M1010)

a. Start Heater.

(1) Move blend air door lever (1) to "HOT" position,

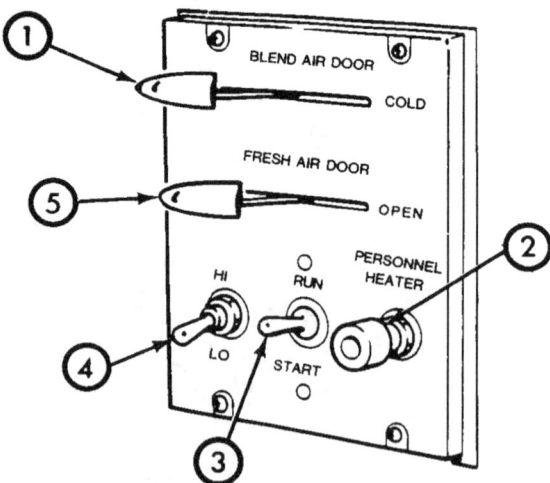

TA466618

2-68

TM 9-2320-289-10

    (2) Move fresh air door lever (5) to "OPEN" position,

    (3) Move HI-LO toggle switch (4) to either "HI" or "LO."

    (4) Hold run-start switch (3) down to "START" position until personnel heater light (2) comes on.

    (5) When personnel heater light (2) comes on, move run-start switch (3) to "RUN."

b. Operate Heater.

    (1) Adjust outlet temperature by moving HI-LO toggle switch (4) to either "HI" or "LO" position.

    (2) Move blend air door lever (1) to any position from "HOT" to "COLD."

    (3) Move tab (7) counterclockwise to open fresh air vent (6) in side wall of patient compartment as desired.

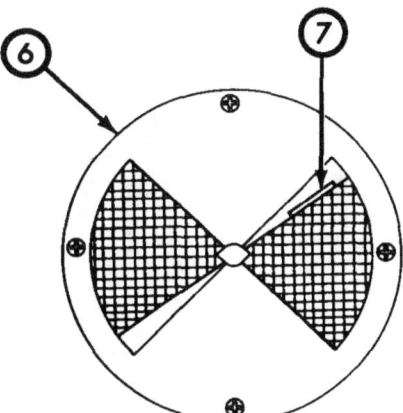

c. Stop Heater.

    (1) Move run-start switch (3) to "OFF."

    (2) Move blend air door lever (1) to "HOT."

    (3) When personnel heater light (2) goes out, move fresh air door lever (5) to "CLOSED."

TM 9-2320-289-10

2-28. OPERATE LITTER BERTHS (M1010)

    a. Three-four Attendant Litter Berth Loading.

NOTE

- Prior to loading e litter patient, litter securing streps must be used to secure patient to the litter. For more information, see FM 8-35.

- Attendant's seat should be stowed forward with seat down to prevent interference with upper litter berth.

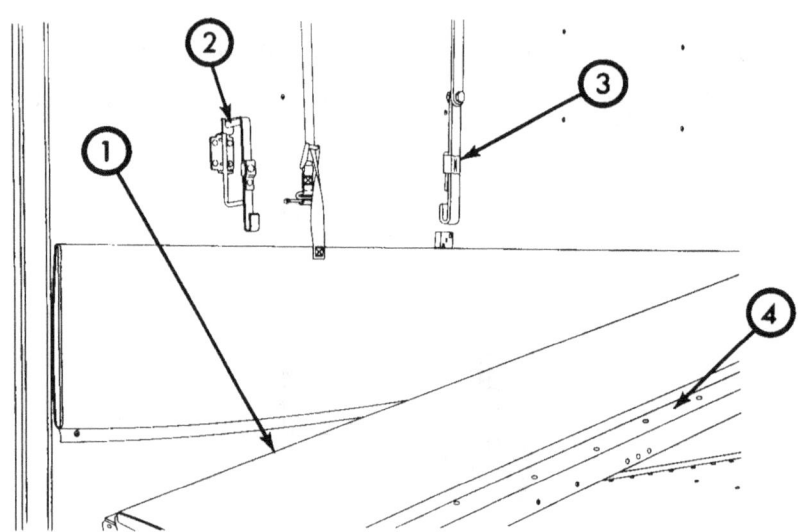

(1) Raise front of one upper litter berth (1).

(2) Place: support pins in holes in forward litter support bracket (5).

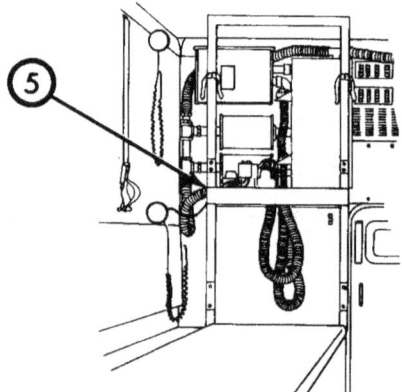

(3) Lift litter and place front stirrups of litter on black groove (4) of inclined litter berth (1).

(4) Slide litter into position onupper litter berth (1).

TA466620

2-70

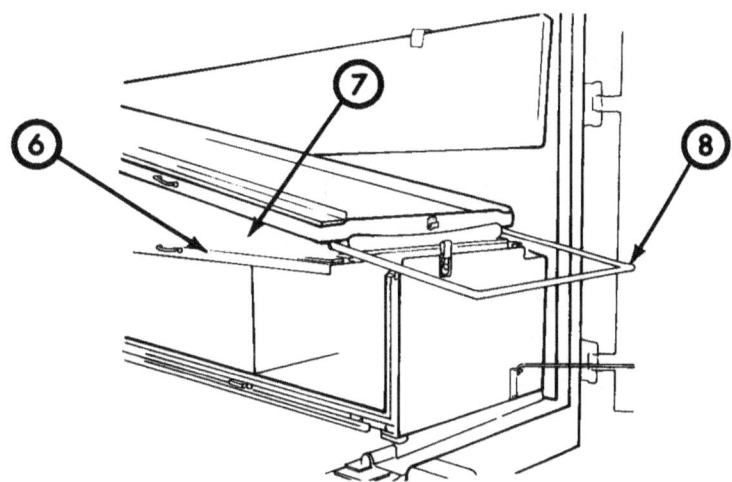

(5) Pull out telescopic berth support rod (8).

(6) Raise end of berth (1) so that berth supports can be attached.

(7) Rotate pivot wall bracket (2).

(8) Push spring clip in ceiling to release hanging bracket (3).

(9) Lower the litter berth (1) into brackets (2 and 3).

(lo) Move locking tabs (14) down.

(11) Slide sleeves13) down over tabs14) to lock bracket (3) into place.

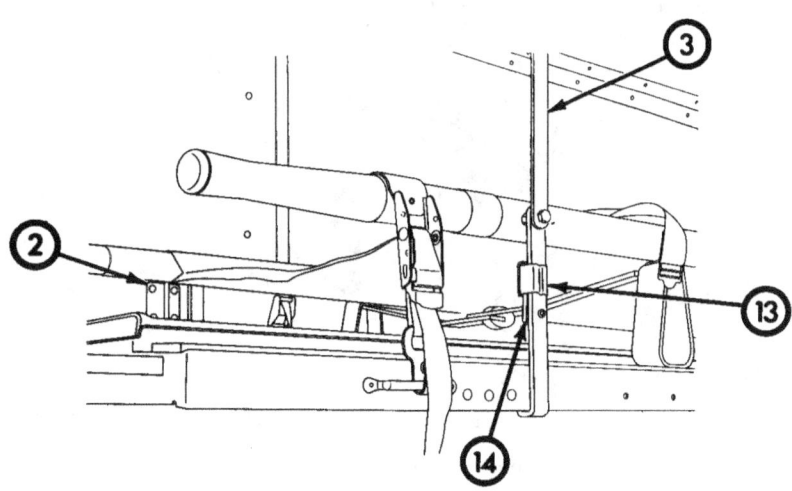

(12) To load lower litter berth (6), lift litter and place front stirrups of litter on black groove (7) of lower berth.

(13) Slide litter completely forward.

TM 9-2320-289-10

b. Two Attendant Loading Method, Using Patient Assist System.

(1) Raise front of one upper litter berth (l).

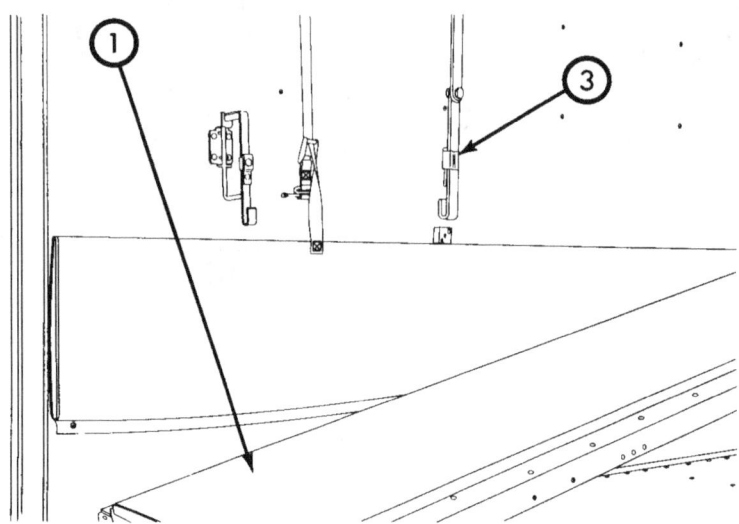

(2) Place support pins in holes in forward litter support bracket (5).

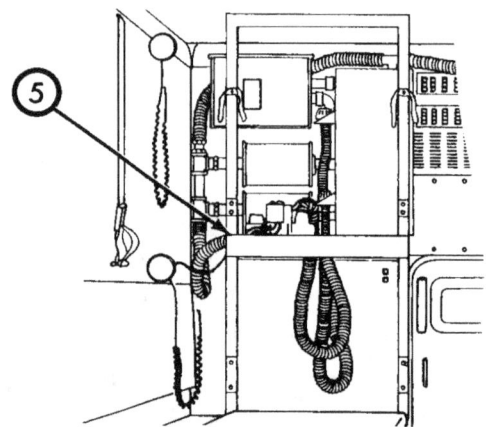

(3) Push spring clip in ceiling.

(4) Pull hanging bracket (3) toward side of truck and down from ceiling, and unfold,

(5) Remove patient assist boom (18) from stowage area.

NOTE

patient assist boom (18) can be mounted on either side of tha truck to load patients.

(6) Mount patient assist boom (18) over rear door on either side of truck.

TA46662 2

2-72

TM 9-2320-289-10

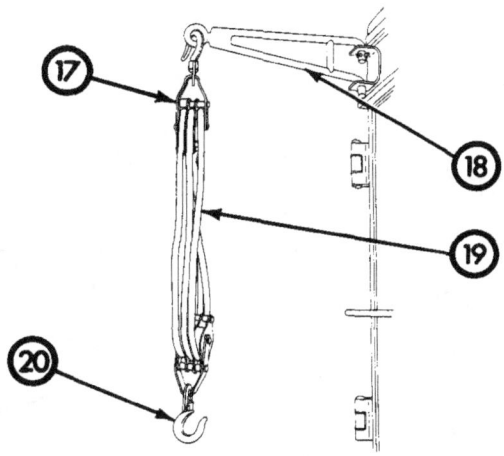

NOTE

When hooking up block and tackle (17) to boom, ensure locking lever is toward the direction of pull end thet ropes ere not twisted.

   (7) Hook block and tackle (17) to patient assist boom (18).

NOTE

The canvas portion of sling can come off the metal eye support for ease of positioning under patient.

   (8) Position litter with patient's chest under hook (20) of block and tackle (17) with patient's head away from truck.

   (9) Slide sling under litter at patient's chest.

   (10) Reassemble sling.

   (11) Hook top of sling to hook (20) of block and tackle (17).

WARNING

Attendant pulling on rope (19) should wear gloves or aerioua rope burna may result.

   (12) One attendant must steady litter while other attendant pulls on rope (19) to raise litter.

   (13) To lock block and tackle (17), pull straight out to side.

   (14) To release block and tackle (17), pull straight down on rope.

NOTE

Top Iitter berth must be Ioaded first.

Guide litter onto litter berth (1) and push litter up berth as far as possible.

One attendant must get into patient compartment to unhook block and tackle (17) and remove sling.

   (15) Both attendants must push litter the rest of the way onto litter berth (1).

   (16)

   (17)

TM 9-2320-289-10

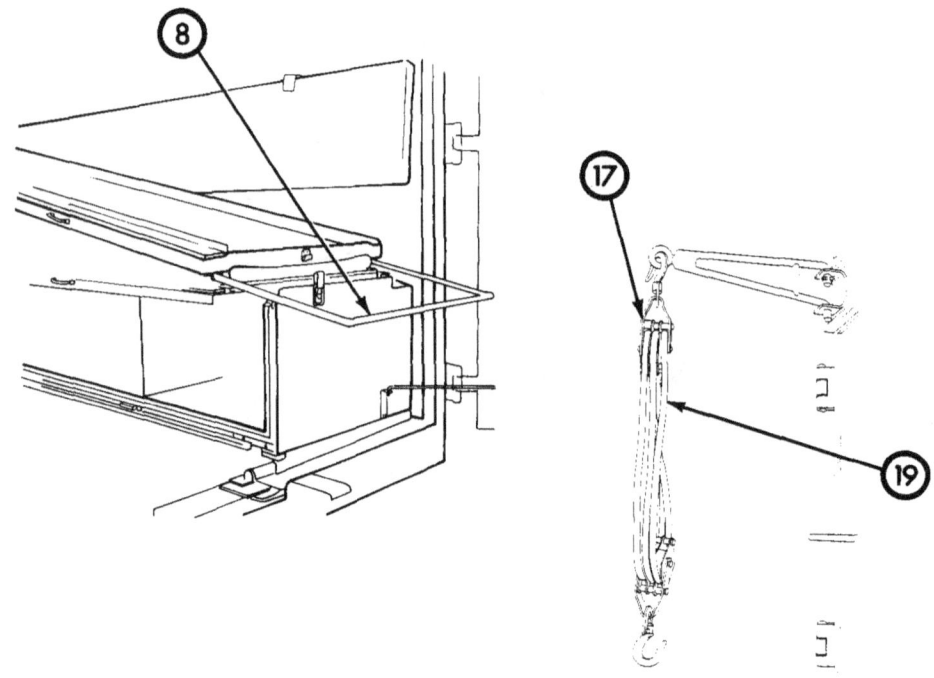

(18) Hook block and tackle (17) to telescopic berth support rod (8)

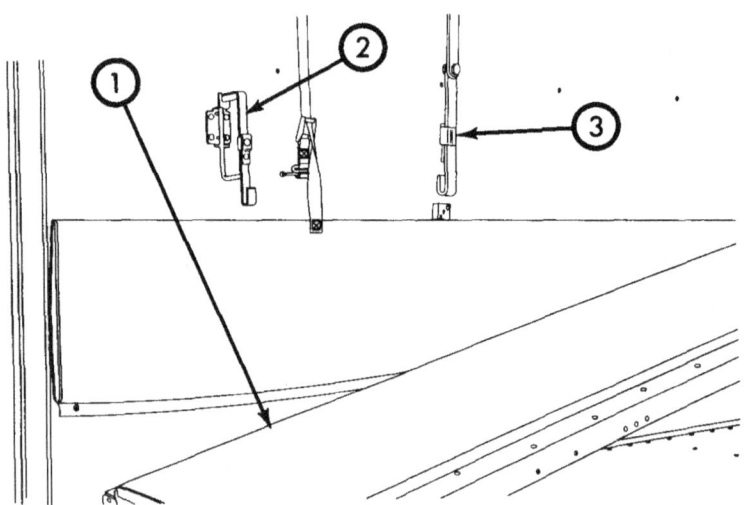

(19) Pull on rope (19) to raise litter berth (1).

(20) Secure litter berth (1) with brackets (2 and 3).

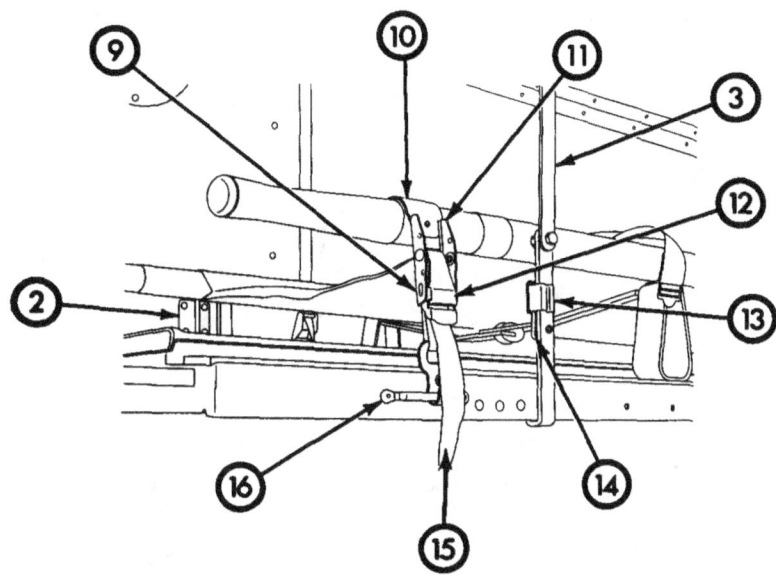

(21) Swing locking tab (14) down.

(22) Put sleeve (13) over tab (14) to hold bracket (3) in place

(23) Use block and tackle (17) to assist in placing patient in bottom litter berth in similar fashion.

c. Litter Tie-down,

(1) Attach tie-down (9) to metal loop (16) in litter berth (1).

(2) Loosen hold-down (11) and strap (15), then relock

(3) Place metal arc end (10) of hold-down (11) over handle of litter and pull strap (15) tight.

(4) Press latch (12) down onto strap (15). Litter is now secure,

(5) To release, pull up on strap (1 5).

TM 9-2320-289-10

2-29. OPERATE ATTENOANT'S SEAT

WARNING

Use extreme caution. To prevent injury, be sure tha seat (2) is locked while the truck is in motion.

a. Lower seat (2) by pushing back on seat with palm of hand and pulling retainer (3) up and out.

b. Push seat (2) down.

NOTE

For ease of movemant, attendant should sit diract front.

c. Sit down on seat (2).

d. Raise or lower seat (2) as needed so that both feet are flat on floor.

e. To raise or lower seat (2), firmly grasp seat and press rearmost pedal (5) at base of seat.

f. Pull up or push down to desired height,

g. Buckle seat belt (1).

h. To move seat (2) forward or backward along track, press frontmost pedal (4) at base of seat.

i. Push with feet.

TA466626

2-30. OPERATE ACCESS STEPS

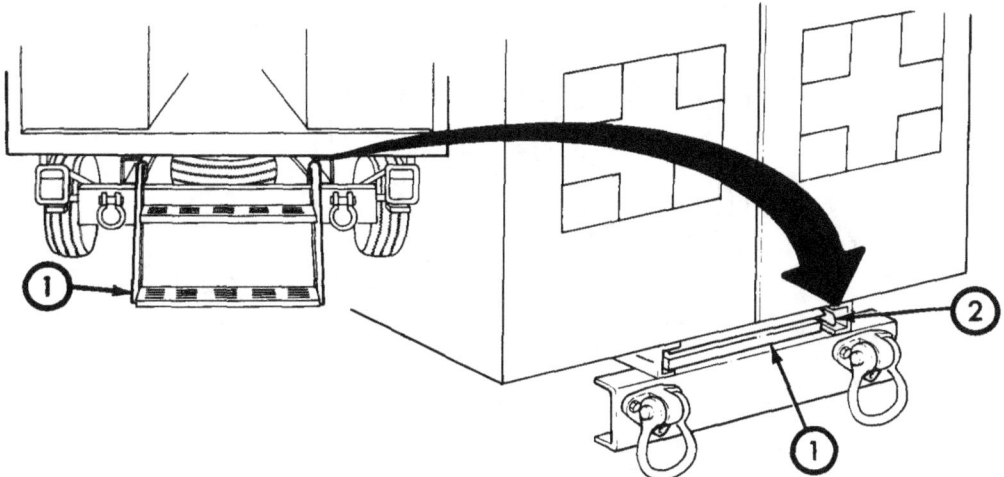

a. Push locking lever (2) on right side of access steps (1).

b. Lift and pull out, then push down on steps (1).

c. Fold steps (1) out of frame.

d. To atow steps, fold steps (1) into frame.

e. Lift up and push steps (1) under frame until locking lever (2) engages.

2-31. OPERATE FOCUS LIGHTS

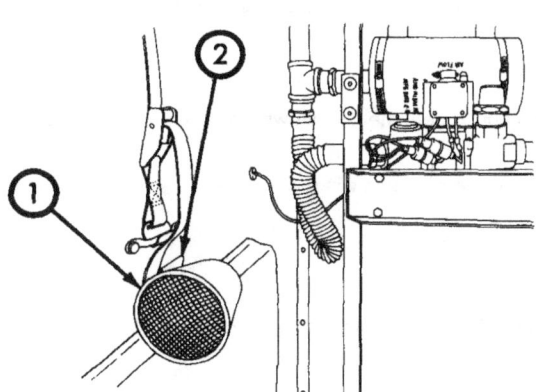

a. To remove focus light (1) from mounting block (2), pull light out.

b. To mount focus light (1), press light into center of mounting block (2).

c. To turn on focus light (1), move small switch at back of light.

d. To turn off focus light (1), move small switch at back of light in opposite direction.

TM 9-2320-289-10

2-32. OPERATE DOMELIGHT

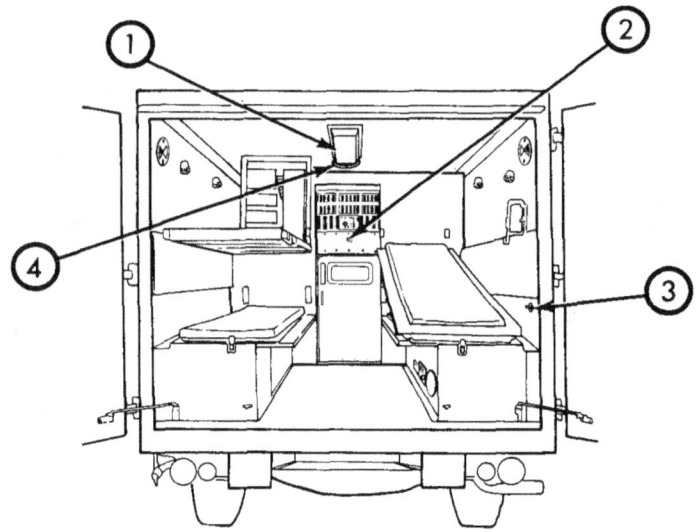

NOTE

Either switch marked "DOME LIGHT" will turn light on or off.

    a. Move domelight switch (3) on right side wall near door or domelight switch (2) under air conditioning unit once to turn on domelight (1) and a second time to turn off domelight.

    b. When service lights/blackout toggle switch is in "BLACKOUT ON" position, and rear doors or door between cab and patient compartment is open, domelight (1) goes off and a blackout light (4) comes on.

    c. If in a blackout condition, make sure all curtains are closed before turning on domelight (1),

2-33. OPERATE BLACKOUT CURTAINS

    a. Lower curtains over all patient compartment windows during hours when blackout conditions are required.

    b. Secure curtains to velcro fasteners around window frames.

2-34. OPERATE SPOTLIGHT

NOTE

Spotlight will not operate in blackout mode.
Move lever (2) to turn on spotlight.

    a. Rotate handle (1) counterclockwise to unlock,

    b. Direct light beam where needed by twisting and rotating handle (1).

    c. Push lever (2) in opposite direction to turn off spotlight when no longer needed.

    d. After operation, rotate handle (1) clockwise to lock.

    e.

TA466628

TM 9-2320-289-10

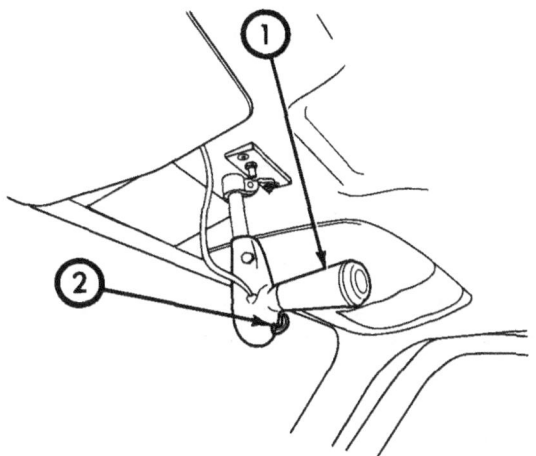

2-35. OPERATE FLOODLIGHTS

NOTE

Floodlights will not operate in blackout mode.

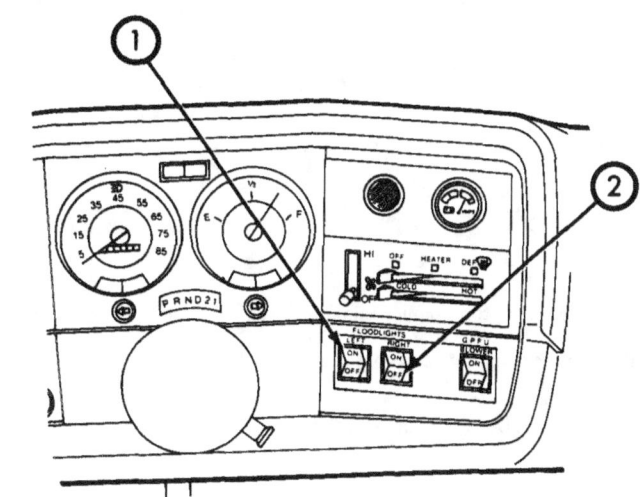

a. Push switch (1) to turn on floodlight on left side of truck.

b. Push switch (2) to turn on floodlight on right side of truck.

c. Push switch (1) to turn off floodlight on left side of truck.

d. Push switch (2) to turn off floodlight on right side of truck.

TM 9-2320-289-10

Section V. OPERATE AUXILIARY EQUIPMENT

2-36. OPERATE WINTERIZATION KIT HEATERS

    a. Operate Underhood Heaters, Below –10°F (–24°C).

      (1) Make sure that ignition key switch is moved to "OFF."

      (2) Pull battery/interior control (7) to "BATTERY."

      (3) Move battery/interior heater hi-low switch (3) to "'HI."

      (4) Hold battery/interior heater run-start switch (4) to "START" until light (2) comes on to show burner is operating (approximately 3 minutes).

      (5) When light (2) comes on, move switch (4) to "RUN. "

NOTE

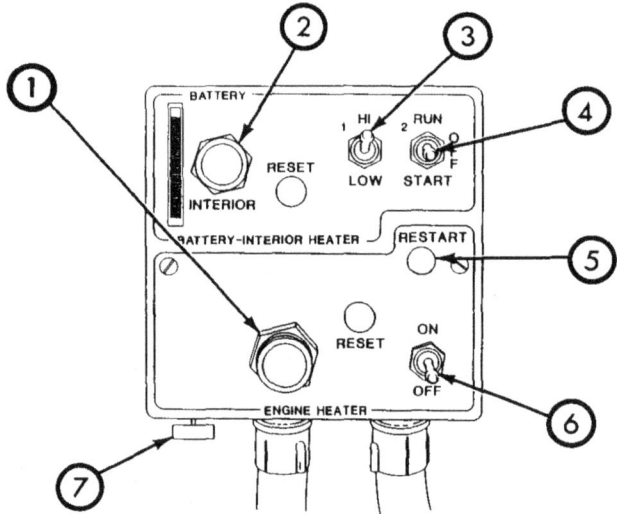

If heater blower does not come on in 5 minutes, heater will shut itself off.

      (6) If heater blower does not come on, move switch (4) to "OFF."

(7) Wait 5 seconds,
Move switch (4) to "START" again.

      (8) Make sure that "RESTART" light (5) comes on.

      (9) Move engine heater switch (6) to "ON. "

      (10) Light (1) will come on in approximately 4 minutes to show burner is operating Wait

      (11) before trying to start engine.

      (12)   (a) Wait forty-five minutes if outside temperature is –50°F (–46°C).

           (b) Wait thirty-five minutes if outside temperature is –40°F (–41°C).

TM 9-2320-289-10

(c) Wait twenty minutes if outside temperature is -30°F (-35°C).

(d) Wait ten minutes if outside temperature is -20°F (-30°C).

(13) Once time period has elapsed, attempt to start engine normally.

(14) After engine is started, move engine heater switch (6) to "OFF."

b. Operate Defroster/Interior Heater.

NOTE

Engine must be running for defroster/interior heating. The battery/interior heater will still be running.

(1) Move battery/interior heater run-start switch (4) to "RUN."

(2) Push battery/interior control (7) to "INTERIOR."

(3) To defrost, move upper heater control lever (6) to "DEF."

(4) Move lower heater control lever (9) to "'COLD."

(5) Move fan speed control (10) to "HI."

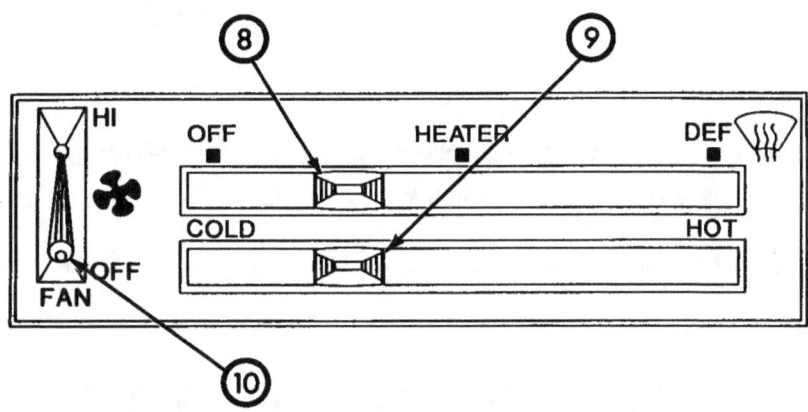

(6) To haat cab, move upper heater control lever (8) to "HEATER."

(7) Move lower heater control lever (9) to maximum "COLD" for best results.

(8) Move battery/interior hi-low switch (3) to either "HI" or "LOW."

(9) Move fan speed control (10) to "HI."

(10) If auxiliary cab heat is not required, move battery/interior run-start switch (4) to "OFF."

(11) For normal cab heating and ventilation, leave battery/interior heater control (7) set to "INTERIOR" and operate regular heater controls.

TA466631

2-81

TM 9-2320-289-10

c. Operate Enclosed Cergo Area Personnel Heater.

NOTE

Engine should be running during heater operation to prevent battery drain.

(1) Move run-start switch (12) to "START" and hold until amber light on heater control panel lights, then move switch to "RUN."

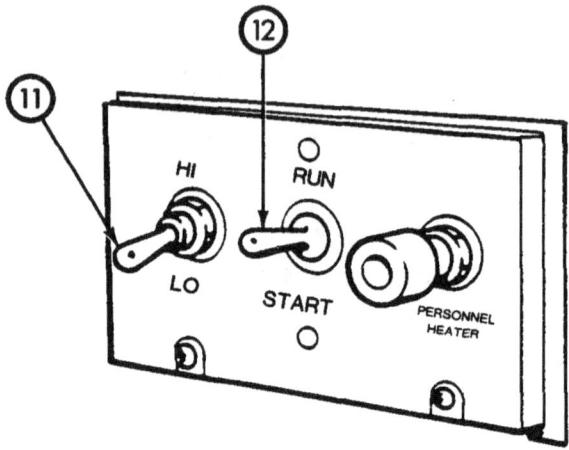

(2) Move HI-LO toggle switch (11) to either "HI" or "LO" for preferred fan operation.

(3) Move air selection lever from "INSIDE AIR" to "OUTSIDE AIR" to mix heated air and fresh air as desired.

(4) When heater is no longer needed, move run-start switch (12) to "OFF." Make sure that amber light goes off.

d. Differences Between Models.

	M1008	M1008A1	M1009	M1010	M1028	M1028A1 M1028A2 M1028A3	M1031
1. Underhood Heaters	X	X	X	X	X	X	X
2. Cargo Area Heater	X	X					
3. Cargo Area Insulator	X	X	X				
4. Hood Insulator	X	X	X	X	X	X	X
5. Cab Insulator	X	X	X	X	X	X	X
6. Plywood Floor Insulator	X	X	X				

TA466632

2-82 Change 5

TM 9-2320-289-10

2-37. CARGO BOX COVER KIT

  a. Stow Cargo Cover.

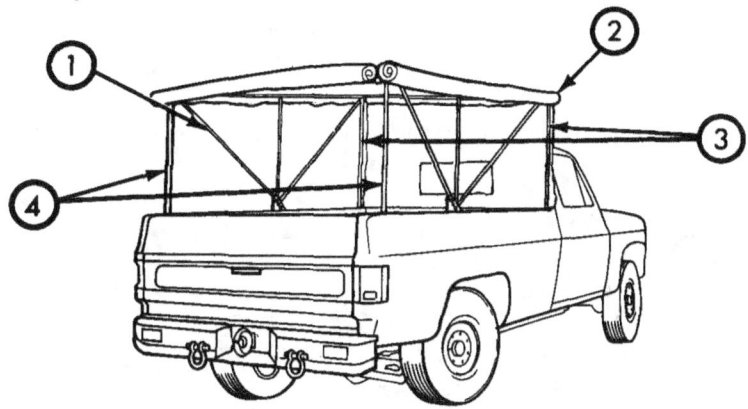

(1) Release front (3) and rear (4) support straps.

(2) From rear of truck, push bows (1) and canvas (2) forward, folding canvas around bows.

(3) Once bows (1) are completely folded down against front of cargo box, secure with atraps.

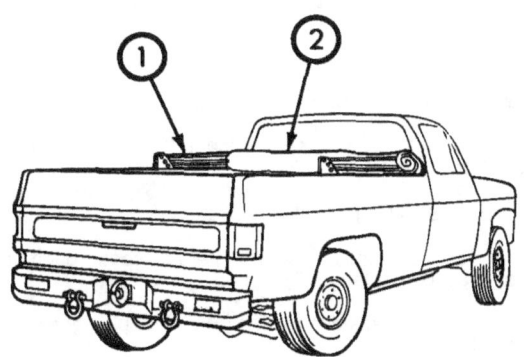

  b. Erect Cargo Box Cover.

  (1) Remove atraps holding canvas (2) and bows (1) together.

  (2) Push bows (1) and canvas (2) toward rear of truck, until canvas is spread evenly and bows are fully extended.

  (3) Secure front (3) and rear (4) retaining straps to brackets on cargo bed wall.

  (4) If situation permits, sides of cover maybe rolled up and secured with strings on inside of cover.

TA466633

2-83

TM 9-2320-289-10

2-38. INSTALL AND REMOVE TROOP SEAT KIT

NOTE

Assistance will be required to perform tha following tasks.

a. Install Troop Saat Kit.

(1) Remove clamps (1) from seat assemblies by pulling hitch pin out of clamp pin hole.

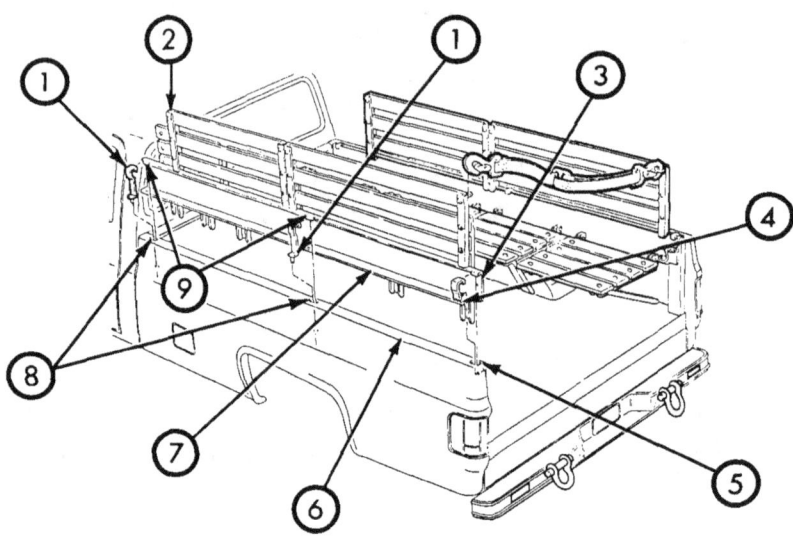

(2) Place a troop seat assembly (2) on top of side rail (6) of cargo box.

(3) Tip seat assembly (3) outward to allow support (9) to fit in stake pockets (8) and to allow lower flange (7) to go beneath pockets.

(4) Place front and middle clamps (1) into stake pockets (8) with clamp hooked over rack.

(5) Ensure that plugs on bottoms of clamps (1) are placed through bottom of stake pockets (8) and holes in bottom of flange (7) on rack.

(6) Place rear clamp (4) in rear stake pocket (5) so that plug goes through bottom of pocket and hole in bottom flange (7).

(7) If required, insert wheel wrench in rear pocket (5). Apply pressure (leverage) against rear clamp (4) and force plug on side of clamp through hole in rear pocket and hole in rack.

Insert hitch pins through holes in plugs on all clamps (l).

(8) Repeat steps for other seat assembly.

(9)

b. Remove Troop Saat Kit.

(1) Put seats in upright position and replace pins.

(2) Remove hitch pin and clamp (1) from all three pockets (5 and 8)

TA466634

2-84

TM 9-2320-289-10

(3) Remove troop seat assemblies (2) and install clamps assemblies.   1) and hitch pins back on seat

(4) Stow seats.

2-39. OPERATE TROOP SEAT KIT

　a. To Stow Seat.

　　(1) Standing in cargo box and facing seat (1), remove quick-release pin and move seat to a stowed position.

　　(2) When seat (1) is in stowed position, fold legs in and place quick-release pin through leg nearest the cab on all seats,

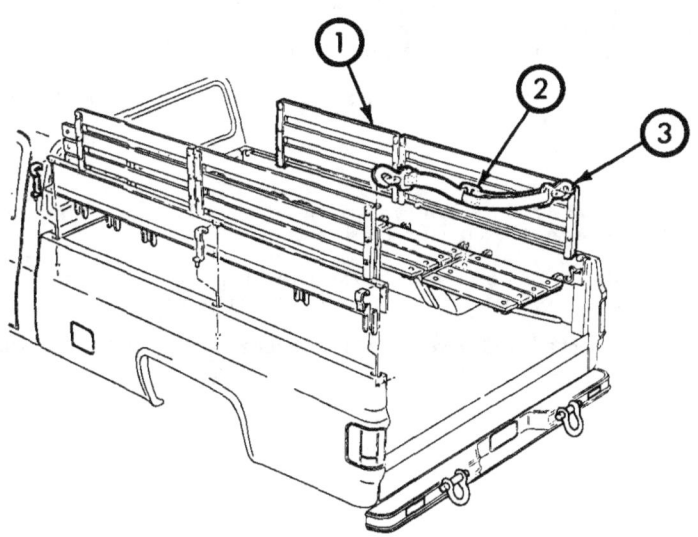

　b. To Place Seat in Down Position.

　　(1) Remove quick-release pin and place legs in angle where truck side and truck floor meet.

　　(2) Place quick-release pin in hole provided in leg under hinge pin,

　　(3) Hook safety strap (2) into eye bolts (3) across back of truck.

TM 9-2320-289-10

Section VI. OPERATE UNDER UNUSUAL CONDITIONS

2-40. SLAVE START TRUCK

WARNING

When slave starting, use NATO slave cables that do not hava loose or missing insulation. DO NOT procead if suitable cables are not available. DO NOT use civilian type jumper cables.

CAUTION

- Wait 3 to 5 minutes after hooking up slave cable to "dead" truck, before attempting to start it. Damage to tha truck's electrical system may result if the truck is started sooner.
- If "dead" truck's engine doas not start within 15 seconds, release ignition key. Wait 3 to 5 minutes before repeating start procedure to pravent overheating the starter and damaging the battaries of the "live" truck. If engina does not start aftar several attempts, organi-zational maintenance must perform addi-tional maintenance.
- Any vehicle with a 24 v system is suitable for slave cable starting. DO NOT attempt to slave start this truck with a 12 v system vehicle.

NOTE

- Before slave starting, make sure checks have been made to determine if the problem is low or dead batteries. If ona battery is missing, DO NOT attempt to slave start.
- If a vehicle other than a CUCV Series truck is used to slave start a CUCV Series truck, refar to the Operator's Manual for that vehicle for any special slave starting procedures.

a. Connect a NATO slave cable to slave receptacle (1) on "dead" truck.

TA466636

2-86

TM 9-2320-289-10

b. cConnect other end of NATO slave cable to slave receptacle on "live" vehicle.

d. Ensure that engine in "live" vehicle is operating.

Start engine of "dead" truck.

Remove slave cable from "live" truck.

e. f. Remove slave cable from "'dead" truck.

g. Stow slave cable.

2-41. TOW DISABLED TRUCK

CAUTION

- NEVER tow by lifting front or rear wheels off ground when truck ie loadad. Stress may cause structural damage.
- A tow bar must be used to prevent damage to truck during towing.
- If truck is to be towed by a wrecker, use only equipment designed for this purpose.
- A safety chain system must be used for all towing. Put safety chain around a portion of truck frame.
- When towing a disabled truck, always shift transmission gearshift Iever to "N" (Neutral), transfer case control lever to "N" (Neutral), unlock front hubs, and release parking brake. DO NOT tow a truck which has become disabled because of damaged transfer case, axle, or transmission. Notify your supervisor.
- NEVER tow a truck at speeds over 35 mph (56 kph).
- NEVER allow passangers to rida in a towed truck for any reason.

a. Tow Truck On All Four Wheels Using Tow Bar.

NOT E

- CUCV Series trucks may be towed on all four wheels with a tow bar, at speeds less than 35 mph (56 kph), for distances up to 50 miles (80 km), providad there is no damage to power train and steering is operable.
- For distances over 5O miles, hava organizational maintenance disconnect rear propshaft at rear axle carrier and front propahaft at front axle carrier, and secure in a safe position.

(1) Remove lifting shackles from front bumper mounting of truck and stow in disabled truck.

(2) Connect legs of tow bar to lifting shackle mounting points on bumper and secure with mounting pins.

NOTE

If towing a disabled CUCV Series truck with a truck that is not CUCV Series, refer to Operator'a Manual for that truck for any spatial procedures during towing.

(3) Connect other end of tow bar to tow pintle of towing truck.

(4) Unlock steering by turning ignition key to "'OFF."

2-87

TM 9-2320-289-10

    (5) Shift transmission gearshift lever to "N" (Neutral).

    (6) Shift transfer case control lever to "N" (Neutral).

    (7) Unlock locking hubs.

    (8) Install safety chains.

    (9) Release parking brake.

NOTE

Remember, power assist for brakes and steering will not be available with engine off.

    (10) Tow truck

  b. Tow Truck on Rear Wheels.

NOTE

For distances over 50 miles, have organizational maintenance disconnect propshaft at rear axle carrier and secure in a safe position.

    (1) Unlock steering by turning ignition key to "OFF,"

    (2) Shift transmission gearshift lever to "N" (Neutral).

    (3) Shift transfer case control lever to "N" (Neutral),

    (4) Unlock locking hubs,

    (5) Release parking brake

  c. Tow Truck on Front Wheels.

NOTE

For distances over 50 miles, have organizational maintenance disconnect propshaft at front axle carrier and secure in a safe position.

- Steering wheel must be firmly secured with wheels in straight-ahead position.

    (1) Unlock steering by turning ignition key to "OFF."

    (2) Shift transmission gearshift lever to "N" (Neutral).

    (3) Shift transfer case control lever to "N" (Neutral).

    (4) Unlock locking hubs.

NOTE

DO NOT use truck's steering column lock or seat belts for clamping steering wheel.

    (5) Firmly secure steering wheel in the straight-ahead position.

    (6) Release parking brake,

2-42. OPERATE IN EXTREME COLD

  a. Truck Start-up.

(1) If truck is equipped with a winterization kit, follow operating instructions contained in paragraph 2-36.

(2) If not equipped with a winterization kit, continue with steps below,

(3) Let engine warm up for at least 5 minutes after it is started.

(4) Drive truck slowly for a short distance as a test run.

(5) Take it easy until truck is warmed up enough to allow normal driving.

(6) Frequently check instruments for any sign of trouble.

(7) If instrument readings are abnormal after engine is warm, IMMEDIATELY SHUT DOWN ENGINE and investigate.

b. At Halt or Parking.

(1) Park truck out of wind if possible.

(2) If shelter is unavailable, park truck facing away from wind.

(39 If truck will be parked for more than a day and high, dry ground is not available:

(a) Make a footing of planks or brush.

' NOTE

DO NOT use parking brake in extremely cold weather. Parking brake cable could freeze.

(b) Shift transmission gearshift lever to "P" (Park) and chock wheels.

(c) Make sure transfer case is in gear.

(d) Clean off any snow, ice, or mud from all parts of truck as soon as possible after parking.

(e) Protect engine and accessories against loose, drifting snow which can melt and then refreeze, causing damage when truck is started up.

(f) Cover truck with canvas if possible.

(g) Check tires with tire gage for proper inflation pressure.

2-43. OPERATE IN EXTREME HEAT

a. Be alert for engine overheating.

b. Engine may overheat during following situations:

(1) Making long, hard pulls in lower gear ranges up steep grades.

(2) Driving in slow, heavy traffic.

(3) Idling for extended periods of time.

(4) Hauling loads close to truck's maximum capacity.

(5) Operating over soft terrain (mud, sand, etc.).

c. To avoid unnecessary overheating conditions:

TM 9-2320-289-10

(1) DO NOT operate with transmission in lower gears.

(2) If engine starts to stall, shift to next lower gear.

(3) Monitor engine coolant temperature light and pull over for cooling off period when necessary.

(4) If any problems develop, such as overheating, refer to Troubleshooting, Table 3-1.

d. At halt or parking:

(1) Park under cover if possible. Direct sunlight will shorten life of rubber, fabric, plastic, and paint.

(2) If cover cannot be found, protect truck with tarpaulins. If entire truck cannot be covered, cover window glass and engine compartment first.

2-44. OPERATE IN UNUSUAL TERRAIN

CAUTION

M1028A2 and M1028A3 rear is wider than front. Use caution in unusual ter-rain, so that flared fender is not damaged.

- DO NOT use "4H" or "4L" trensfer case ranges on dry, hard-surfaced roads or the transfer case will bind, tires will wear excessively, and damage to truck may result.

NOTE

Four-wheel drive is used to provide additional traction and lower gear for use in off-road operations and to provide low-speed pulling power in unusual conditions. The four-wheel ("4L" and "4H") ranges of the transfer case should be used ONLY when greater traction and power are required in off-road operations.

a. General Guidelines. Use good judgement when driving off-road over rough or unusual terrain. Follow these guidelines:

(1) Keep engine at moderate speeds.

(2) DO NOT let wheels start spinning. If wheels start to spin, ease off accelerator pedal and attempt to regain traction.

(3) DO NOT let air out of tires in sand to gain traction.

b. Use of "4H" and "4L" Transfer Case Range. Use four-wheel drive ONLY on soft ground or when driving on ice, through deep, loose snow, deep sand and deep mud, and where maximum traction is a must.

c. Driving Techniques in Four-wheel Drive.

NOTE

A truck in four-wheel drive will accelerate much faster than a two-wheel drive truck on snow, but will not stop any more quickly.

Locking hubs must be in the "LOCK" position.

- (1) Before climbing a steep grade, shift transfer case control lever to "4L" and transmission gearshift lever to "1." If wheels start to slip, "walk" the truck the last few remaining feet of a hill by swinging front wheels sharply left and right if situation permits. This action will provide fresh "bite" into the surface and will usually result in enough traction to complete climb,

- (2) You can proceed safely down a grade in four-wheel drive by shifting transfer case control lever to "4L" and transmission gearshift lever to "1." Let the truck go slowly down the hill with all four wheels turning against engine compression.

2-90 Change 5

(3) When moving across e slope, choose the least angle possible, keep moving, and avoid turning quickly.

WARNING

If you go into a skid, DO NOT USE BRAKES until you ara straightened out. It will only cause greater loss of control.

d. Snowy or Icy Roada and Terrain. If rear end skidding occurs:

(1) Turn steering wheel in direction of skid.

(2) Let up on accelerator pedal and apply service brake pedal in a gradual, pumping manner.

2-45. FORDING

CAUTION

Never attempt to ford e crossing deeper than 20 inches (51 cm). The truck's components may be dameged.

Determine feasibility of fording by making sure bottom of crossing is not too soft to support truck.

a.

b. Shift transmission gearshift lever to "1."

c. Shift transfer case control lever to "4L."

d. Slowly enter water, but don't let engine stall.

e. Limit speed to 5 mph (8 kph).

f. After fording, don't rely on brakes until they have been tested.

g. Depress brake pedal with light pressure several times while moving truck. This will dry out brakes.

h. Do after fording maintenance. Refer to paragraph 3-11.

TM 9-2320-289-10

2-46. OPERATE SWINGFIRE HEATER

    a. Prepare Swingfire Heater for Operation.

        (1) Remove fuel tank cap (1). Fill heater fuel tank (5) with clean gasoline, and install and tighten fuel tank cap.

        (2) Check operation of air shut-off valve (2) by pushing pressure pin (3). Pressure pin must bounce back.

        (3) Ensure that diaphragm valve (4) is in the "+" position

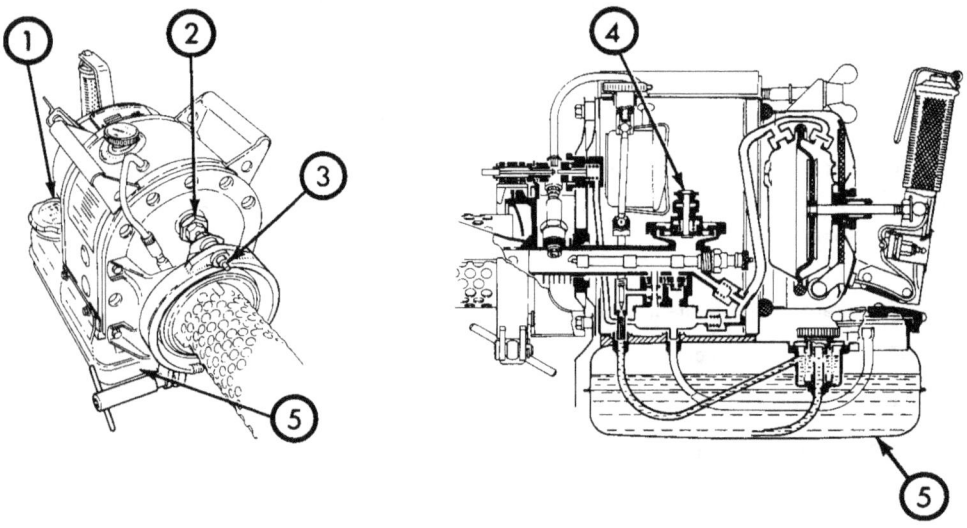

    b. Start Swingfire Heater.

WARNING

DO NOT operate heater in enclosed areas. Exhaust gases could cause injury or death.

        (1) Plug starting cable (10) connector into receptacle on left front fender of truck.

        (2) Plug other end of starting cable (10) into heater handle (1 1).

        (3) Squeeze pump lever (7), which pushes on pushbutton switch (6).

        (4) Listen for humming sound. If humming sound is not heard, go back to step (1) and repeat procedure. If humming is heard, continue.

NOTE

If during starting procedure a thick white fuel fog comes from pulsation pipe and pulsation stops, heater is flooded. (See Correction of Flooded Condition, below.)

        (5) Turn fuel regulator knob (8) fully counterclockwise (Closed).

2-92

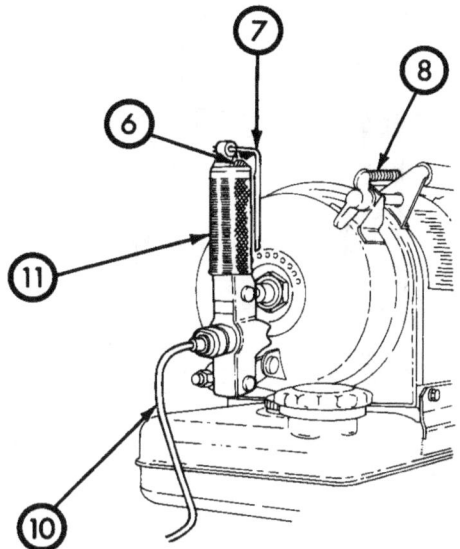

NOTE

DO NOT pump heater handle (11) while depressing pushbutton switch (6).

(6) If temperature is above 0°F (-18°C) go on to step (7). If temperature is below 0°F (-18°C) squeeze pump lever (7) to depress pushbutton switch (6) to activate preheating plug and hold down for the following time periods:

   (a) 0°F to -20°F (-18°C to -29°C) depress for 2 minutes.

   (b) -20°F to -30°F (-29°C to -30°C) depress for 3 minutes.

   (c) -30°F to -40°F (-34°C to -40°C) depress for 4 minutes.

   (d) -40°F to -50°F (-40°C to -46°C) depress for 5 minutes.

(7) Move pump lever (7) forward and backwards 3 or 4 times.

(8) With one hand, turn fuel regulator knob (8) clockwise (Open) 1/2 to 1 turn if temperature is above 0°F (-18°C), or 1 to 1-1/2 turns if temperature is below 0°F (18°C), while continuing to operate pump lever (7) with other hand.

(9) When first pulsating sounds are heard, continue pumping and adjusting fuel regulator knob (8) until pulsating sounds come at regular intervals.

(10) At this point, stop pumping; the heater has been started.

(11) Allow heater to operate for 3 to 5 minutes while fine adjusting the fuel/air mixture using the fuel regulator knob (8).

(12) Once heater is properly operating, disconnect the starting cable (10) from the heater and from the truck, and stow the cable.

TM 9-2320-289-10

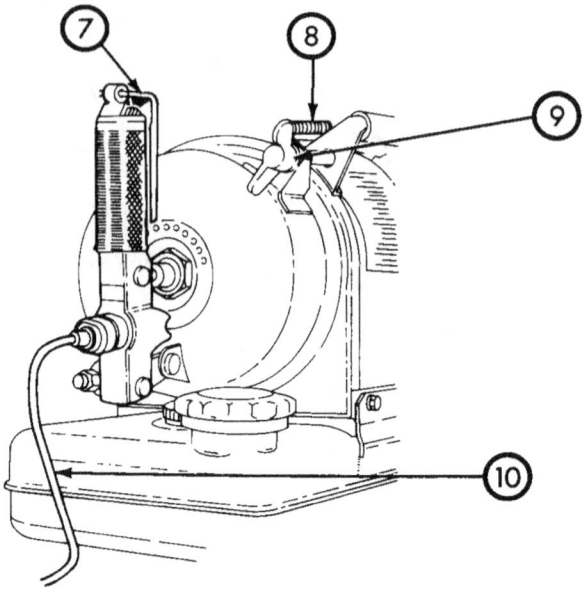

c. Correction of Flooded Condition,

NOTE

Dense white fuel fog is the indication that tha heater has flooded.

(1) Close fuel regulator knob (8),

(2) Operate pump lever (7) until surplus fuel has been blown out exhaust pipe and pulsations start again,

(3) Once pulsations start again, open fuel regulator knob (8) 1/2 to 2 turns and fine adjust the fuel/air mixture.

(4) If heater has flooded to point that fuel overflows the diaphragm valve into the mixing chamber, perform the following steps:

    (a) Disconnect starting cable (10) from truck receptacle,

    (b) Unscrew wingnut (9).

    (c) Open chamber cover.

    (d) Tilt heater to the left to let fuel run out.

    (e) Clean up runoff fuel and allow chamber to dry.

    (f) Restart heater.

2-47. START TRUCK WITH AI13 OF SWINGFIRE HEATER

a. When heater is operating, and starting cable is removed, open clamp by turning turn buckle (1) counterclockwise,

b. Position heater in water jacket, at left front fender of truck, and rotate turnbuckle (1) 16 full turns to lock heater clamp to water jacket.

TA466639

2-94

TM 9-2320-289-10

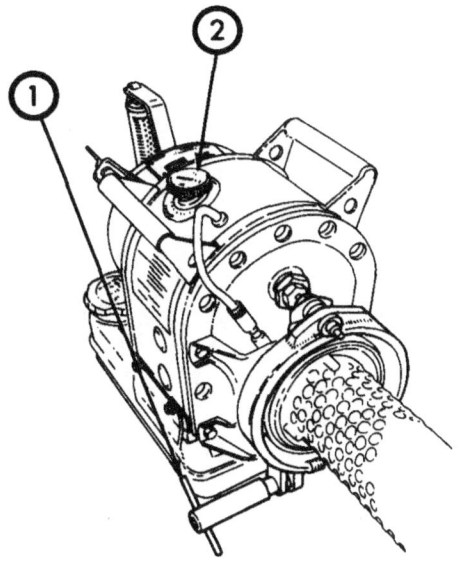

c. Allow heater to operate until frost melts from engine block.

d. Start engine in normal manner

e. With engine running, turn off heater by completely closing fuel regulator knob (2).

f. Remove heater from water jacket.

NOTE

If heater is not to be used for 48 hours or longer, drain all fuel from fuel tank.

g. Allow heater to cool before stowing.

TM 9-2320-289-10

CHAPTER 3

MAINTENANCE INSTRUCTIONS

Section I. LUBRICATION

3-1. GENERAL

NOTE

Lubrication procedures ere performed et the organizational maintenance level.
You may be required to assist the mechanic.

Periodic servicing ensures that the truck will operate at peak performance. Lubrication Order LO 9-2320-289-12 gives complete cleaning and lubricating instructions. Refer to "NOTES" for specific instructions on lubrication. Service intervals are based on normal operation under normal conditions.

3-2. LUBRICATION ORDER LO 9-2320-289-12

A copy of the Lubrication Order (LO) is issued with each truck and must remain with it at all times. If you receive the truck without a copy, immediately notify your supervisor.

REMEMBER:

a. The time to change oil is when starting or other operations become sluggish, or when outside temperatures move out of the appropriate range for the type of oil currently in the truck. Do not wait for the next normally scheduled oil change.

b. When you are operating under extreme conditions, lubricants should always be changed more frequently than the normal intervals specified by the LO. Lubricants break down or become contaminated more frequently under extreme conditions.

TM 9-2320-289-10

Section II. TROUBLESHOOTING

3-3. SCOPE

    a. This section contains troubleshooting information that will help you locate and correct operating problems. Each problem is followed by a list of tests or inspections. These will help you to determine probable causes and corrective actions in the order listed.

    b. Not all malfunctions that mayoccur are listed. All tests or inspections and corrective actions are not listed. If a malfunction is not listed or is not corrected after you take the listed actions, return to organizational maintenance, if you can. Otherwise, stop the truck and notify your supervisor.

3-4. TROUBLESHOOTING

<u>CAUTION</u>

When you get the truck running again, IMMEDIATELY RETURN to organizational maintenance. Corrective action is designed only to let you return to motor pool.
Continued operations could damage the truck.

    a. Basic Procedure. If you have trouble starting the truck or have trouble on the road, some simple repairs can be made to get the truck running again. First, identify the MALFUNCTION (or problem). Then locate it in Table 3-1 and follow the directions in the CORRECTIVE ACTION column.

    b. If You Cannot Fix It. If you cannot pinpoint the problem, or if it is too serious to repair, stop the truck and notify your supervisor at once.

TM 9-2320-289-10

3-5. SYMPTOM INDEX

                                                                                               Troubleshooting
                                                                                                  Procedure
                                                                                                    Page

ELECTRICAL SYSTEM

    Gages:
        Inoperative   3-7
        Operating abnormally   3-7
    Horn:
        Inoperative.   3-7
        Operates continually   3-7

ENGINE

    Lack of Power   3-6
    Lubricant Leaks 3-7
    Misfires 3-7
    Overheats . . . . .3-6 Will
    Not
        Continue to run   3-5
        Idle   3-7
        Start, but turns over 3-4
        Turnover 3-4

TIRES

    Abnormal Wear   3-8

TRANSFER CASE

    Tranafer Case Remains in Four-wheel Drive   3-8

TM 9-2320-289-10

Table 3-1. Troubleshooting

MALFUNCTION
    TEST OR INSPECTION
        CORRECTIVE ACTION
---

1. ENGINE WILL NOT TURN OVER WHEN IGNITION SWITCH IS TURNED TO THE START POSITION

CAUTION

Use only NATO slave cables and slave receptacle mounted on front to slave start the truck. DO NOT attempt to use civilian style jumper cables to slave start truck. Damage can occur to both "dead" and "live" trucks if slave procedures and this caution are not followed.

  Step 1. Check position of transmission gearshift lever.

        Shift transmission gearshift lever to "P" (Park) or "N" (Neutral).

  Step 2. See if battery connections are loose

        Notify your supervisor.

  Step 3. Battery may be discharged. Look at charge indicator on battery.

        Notify your supervisor if battery is discharged.

| DARKENED | DARKENED | LIGHT INDICA- |
| TOR | INDICATOR | YELLOW OR |

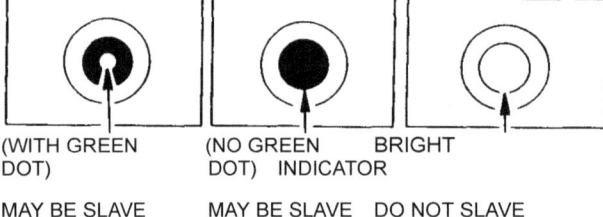

| (WITH GREEN DOT) | (NO GREEN DOT) | BRIGHT INDICATOR |
| MAY BE SLAVE STARTED | MAY BE SLAVE STARTED | DO NOT SLAVE START |

2. ENGINE TURNS OVER BUT WILL NOT START

    Step 1. Fuel tank may be empty. Check gage with ignition switch in "ON" position.

        If empty, fill tank.

NOTE

DO NOT put fuel in intake manifold or use starting fluids.

Assistant will be
- required for the following procedures.
-

TM 9-2320-289-10

Table 3-1. Troubleshooting - Continued

MALFUNCTION
TEST OR INSPECTION
CORRECTIVE ACTION

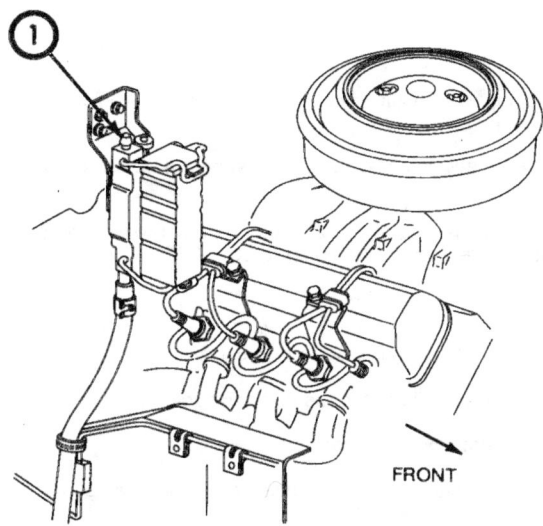

Step 2. Open air bleed valve (1) on top of fuel filter on front of firewall

Step 3. Crank engine at 10-15 second intervals after WAIT light goes out, Continue until fuel flows out of air bleed valve (1).

Step 4. Close air bleed valve (1).

Step 5. Crank engine for 10 seconds. If it does not start, wait 15 seconds and crank engine again after WAIT light goes out. Refer to paragraph 2-8 for proper starting procedures.

Step 6. If after three attempts, engine does not start, notify your supervisor.

Step 7. Battery may be weak.

    Use same procedures outlined in Malfunction 1, Step 3.

3. ENGINE STARTS BUT DOES NOT CONTINUE TO RUN

Step 1. Fuel level may be low. Check fuel gage with ignition switch in the "ON" position.

    Add fuel if necessary.

Step 2. Check for water in fuel.

    Drain fuel filter following procedures outlined in paragraph 3-7. Restart engine. Refer to paragraph 2-8.

TA466642

TM 9-2320-289-10

Table 3-1. Troubleshooting - Continued

MALFUNCTION
    TEST OR INSPECTION
        CORRECTIVE ACTION

4. ENGINE OVERHEATING

WARNING

NEVER remove radiator cap when engine is hot. This is a pressurized cooling system and escaping steam or hot water can cause serious burns.

Step 1. Check coolant level in see-through recovery tank. Observe low coolant warning light to see if it is on.

    If coolant level is low, obtain coolant mixture from organizational mainte-nance and put it in recovery tank. Refer to LO 9-2320-289-12.

CAUTION

DO NOT bend radiator cooling fins.

Step 2. Look at outside of radiator for obstructions such as leaves, etc. Remove any-

thing that blocks the core or impedes air flow.

Step 3. Check cooling system hoses for evidence of leaks (wet spots, escaping steam, or water).

Notify your supervisor. Check

engine oil level.

Step 4. Refer to LO 9-2320-289-12.

WARNING

NEVER remove radiator cap when engine is hot. This is a pressurized cooling system and escaping steam or hot water can cause serious burns.

CAUTION

Kaep a careful eye on oil pressure and engine coolant temperature lights when you are driving a truck that you know is low on oil or coolant. You may still be losing oil or coolant. If either light comes on, IMMEDI-ATELY SHUT DOWN ENGINE and do not mova truck any further.

Step 5. After engine cools down, check coolant level in radiator. Radiator should be full.

    If coolant level is low, obtain coolant mixture from organizational main-tenance and put it in radiator. Refer to LO 9-2320-289-12.

5. LACK OF POWER

    Step 1. Your parking brake may be applied.

        Release your parking brake.

    Step 2. Recheck your load. You could be overloaded.

        Lighten load, if possible, or use a lower driving range.

TM 9-2320-289-10

Table 3-1. Troubleshooting - Continued

MALFUNCTION
    TEST OR INSPECTION
        CORRECTIVE ACTION

6. ENGINE WILL NOT IDLE OR IT MISFIRES

<u>WARNING</u>

Care should be taken when working eround engine menifold. Hands or arms could be burned.

    Step 1. Check for water in fuel,

        Drain fuel filter. Refer to paragraph 3-7.

7. GAGES INOPERATIVE OR OPERATING ABNORMALLY

    Step 1. Turn ignition switch to "ON" and start engine. Generator and oil pressure lights should immediately go out. Observe lights. (See paragraph 2-2 for normal readings)

        If lights register abnormally, SHUT DOWN engine.

CAUTION

DO NOT drive truck with a high oil level. Too much oil can blow out oil seals damage main bearings, and cause other damage requiring a complete ovarhaul.

    Step 2. Oil pressure light stays on. If light stays on when engine is warm and you are not using engine compression for braking, engine may be overfilled with oil.

        If possible, park truck on level ground. Wait two minutes and check engine oil level. If it is over the "FULL" mark on dipstick, notify your supervisor.

8. LUBRICANT LEAKS

    Step 1. Observe drain plugs for leaks and check oil level.

        If plugs are loose or oil level on dipstick is low, notify your supervisor. Loss of

    Step 2. engine oil may be caused by oil level being too high.

        Notify your supervisor.

9. HORN INOPERATIVE OR OPERATES CONTINUALLY

NOTE

If service lights/blackout toggle switch is at "ALL OFF," horn will not work.

    Step 1. Move service lights/blackout toggle switch to "SERVICE LIGHTS ON." If horn does not operate when you push the horn button:

        Notify your supervisor.

    Step 2. When horn operates continually or operates occasionally without button being depressed:

        Notify your supervisor.

TM 9-2320-289-10

Table 3-1. Troubleshooting - Continued

MALFUNCTION
    TEST OR INSPECTION
        CORRECTIVE ACTION

1. ABNORMAL TIRE WEAR

    Step 1. Check for proper tire pressure Inflate low tires to

    proper pressure

NOTE

Maka sure "4H" and "4L" positions in transfer case are not used on hard-surfaced roads. Use "4H" and "4L" positions in transfer case only when maximum traction is neaded. Refer to paragraph 2-44.

    Step 2. Truck alinement may not meetspecifications.

    Notify your supervisor.

1. TRANSFER CASE REMAINS IN FOUR-WHEEL DRIVE

    Step 1. Shift transmission gearshift lever to "R" (Reverse) and drive in reverse for 15 feet (4.6m) while attempting to shift transfer case control lever to "2H."

    Step 2. Drive forward and ensure that truck has been taken out of four-wheel drive. If not, repeat step 1.

    If truck will not come out of four-wheel drive after two attempts, notify your supervisor.

TM 9-2320-289-10

Section III. MAINTENANCE PROCEDURES

3-6. CHANGING TIRES

NOTE

- The jack on tha M1009 is located under the front passenger seat; behind tha driver's seat on the M1010; and bahind the bench seat on the right-hand side of the cab for all other trucks.

- The truck has e spare tire Iocated under the frame, in the rear of the truck, on all trucks except the M1009, The spare tire on the M1009 is located behind the raar eeat on the right side of the truck.

a. If possible, park on a level surface. Set parking brake tightly.

b. Shift transmission gearshift lever to "P" (Park) and shift transfer case control lever to "2H," "4H," or "4 L."

c. Turn on hazard warning flasher (if tactical situation permits).

WARNING

Keep clear of spare tire when removing or replacing. It could cause injury to personnel.

d. Remove spare tire and jacking tools from stowage areas.

M100 9

TM 9-2320-289-10

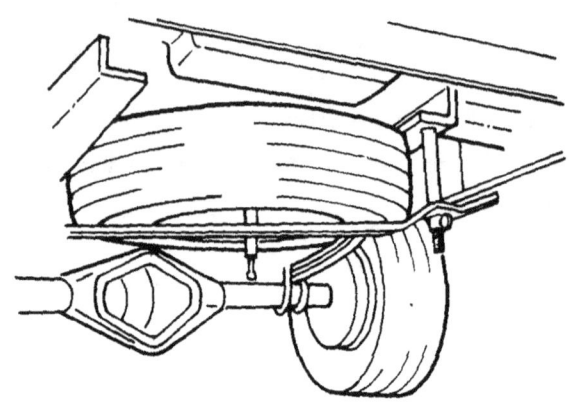

All Except M1009

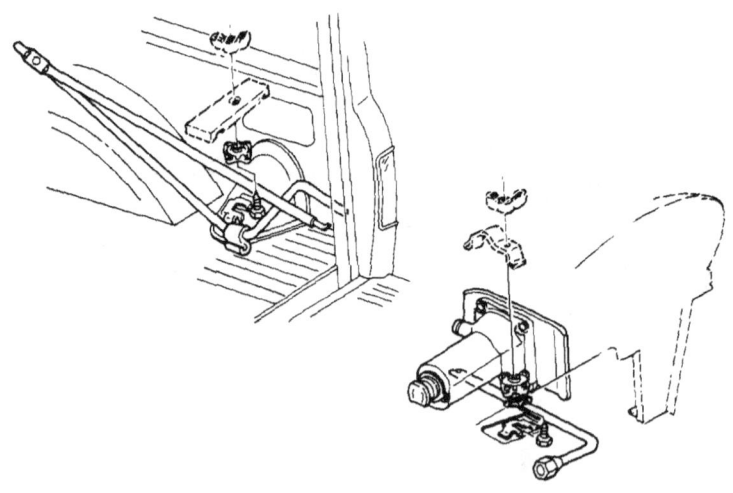

M1009

TM 9-2320-289-10

All Except M1009

e. Chock the front and rear wheels located diagonally opposite the one you are changing.

f. Raise the jack to just below axle before positioning by turning jack handle with a slow, smooth motion.

WARNING

Do not operate the engine and make sure all personnel are outside of the truck when the truck is raised on the jack. Serious injury can result if the truck moves and falls off the jack.

g. Position jack under front axle as close to wheel as possible. On rear axle, position jack between "U" bolts. Raise jack until tension is on axle. Loosen wheel lug nuts about one-quarter turn.

NOTE
Perform steps h and I for all except M1028A2 and M1028A3.

h. Raise wheel high enough off the ground, so that an inflated tire will just clear the surface. Remove lug nuts and then remove wheel.

i. Mount spare tire and finger tighten lug nuts.

TM 9-2350-253-10

NOTE

Perform steps 1.1 and 1.5 for M1028A2 and M1028A3 only.

i.1 Raise wheel(s) high enough off the ground, so that an inflated tire will just clear the surface,

i.2 Remove lug nuts, clamping plate and remove outer rear or front wheel.

NOTE

Perform steps i.3 and i.4 only if inner rear wheel must be re-moved.

Remove spacer plate and inner rear wheel,

i.3

CAUTION

If spacer plate is left out, lug nuts cannot be tightened properly, and damage to vehicle could occur.

Install spare wheel and spacer p!ate on rear hub.

i.4 NOTE

When installing outer rear wheel, ensure that valve stems are positioned 180° apart, and that the inner valve stem is accessi-ble through a hole in the outer rim.

Install outer rear or front wheel, clamping plate and finger tighten lug nuts.

i.5

TM 9-2320-289-10

NOTE

Proper torque for Ml 009 is 90 Ib.-ft. (120 N.m). Proper torque for all others is 140 Ib.-ft. (190 N.m).

j. Lower the truck by turning jack handle counterclockwise. Retighten lug nuts in sequence shown. Have organizational maintenance apply proper torque as soon as possible after changing the spare tire.

k. Properly stow the tire and jack equipment,

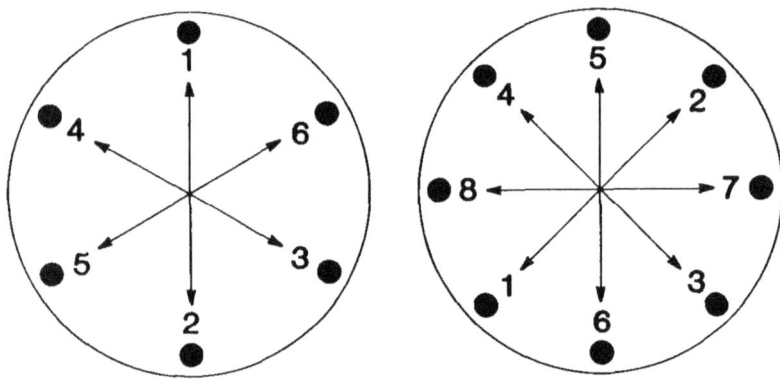

l. Remove and stow chocks.

m. Release parking brake before driving 3-7.

DRAIN FUEL FILTER

WARNING

Do not work on a hot engine. Allow engine to cool to a safa temperature. Failure to observe this warning can result in injury to personnel.

NOTE

You will need an assistant to perform the following procedure.

a. Drain water from fuel filter

(1) With engine off, parking brake applied, and gearshift lever in "P" (Park), open the hood. Place a suitable container under the fuel filter water drain hose (1) located at the rear of the right front wheelwell by the spring shackle.

TM 9-2320-289-10

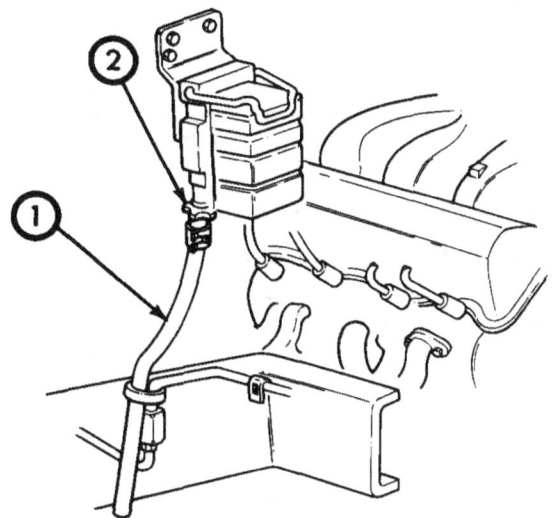

(2) Remove truck fuel tank cap.

NOTE

DO NOT tamper with tha air bleed screw located on top of fuel filter.

(3) Open the water drain petcock (2) on the bottom of fuel filter by turning counterclockwise until water drains from hose (1).

(4) Have an assistant observe drain hose (1).

(5) Start the engine and allow it to idle while water drains f rom the hose (1) into container.

NOTE

If clean/clear fuel doaa not become apparent, notify your supervisor.

(6) Have assistant signal when all evidence of water is gone. One to two minutes should be long enough.

(7) Shut off engine and close petcock (2) by turning clockwise. Do not use force or tools. If fuel continues to leak, notify your supervisor.

(6) Replace fuel tank cap.

b. Restart engine and let run fortwoto three minutes. Engine may run rough for a few minutes. If it stops and will not restart, notify your supervisor. If WATER-IN-FUEL light comes on again after a short period of time, notify your supervisor.

TM 9-2320-289-10

Section IV. MAINTENANCE UNDER UNUSUAL CONDITIONS

3-8. EXTREME COLD WEATHER

    a. Preparing the Truck. The truck requires careful preparation when it is to be operated in extreme cold weather. Extreme cold, below 0°F (-18°C) can thicken lubricants, cause electrical short circuits by cracking the insulation, and make structural materials hard, brittle, and easily damaged. In general, when preparing a truck for extreme cold weather service, follow these rules:

        (1) Schedule truck for cold weather servicing.

        (2) Make sure cooling system is protected. Refer to FM 9-207 for mandatory instructions.

    b. Cold Weather Maintenance Requirements. When temperature drops below 0°F(-18°C), be especially careful while you do your PMCS. Always report any operating problems or trouble in starting to your supervisor.

WARNING

    Do not let solvents come in contact with your skin in extreme cold weether. Rapid evaporation causes supercooling and can give you e serious case of frostbite within minutes.

    Always allow extra tima for warm-up during this kind of weather. In arctic temperatures as low as -50°F(-47°C), it may take up to two hours just to reach normal operating ranges.

    c. For Lena Halts or Extended Shutdown. Park the truck in a warm place. if possible. You do not have to drain out the oil since it will remain fluid.

3-9.   EXTREME HOT OR HUMID WEATHER

    a. Hot Weather. In hot weather, always keep a watch on coolant levels. Don't let recovery tank get low. Be especially careful to keep radiator free of leaves or other debris. If coolant level drops, look for leaks in the cooling system. When checking your tire pressures with a tire gage, make sure tires are not overinflated. DO NOT check tires immediately after running on the road. Wait until tires have had a chance to cool off. Have the cooling system, oil filter, and air cleaner checked and serviced frequently.

    b. Humid Weather. In hot, damp climates be watchful for accelerated corrosion. Evidence of deteriorating materials includes rust and paint blisters on metal surfaces, and fungus, mold, or mildew on fabrics and glass. Check the truck daily and act on these problems as soon as you notice them. If you find rusting metal or fungus, clean affected area and apply a thin coat of light oil.

3-10. AFTER OPERATING ON UNUSUAL TERRAIN

CAUTION

    DO NOT hose off radiator with a high pressure hose. High pressure can bend radiator fins and reduce the radiator's cooling ability. (See paragraph 2-6)

    DO NOT hose out Inside of cab, Water can accumulate and cause rust, or can cause damage to electrical components in cab.

    a. Mud. Clean off truck as soon as possible after operating in mud. Pay careful attention to the radiator cooling fins. Hose off the radiator if you find mud deposits. Check air filter and have it replaced if necessary.

TM 9-2320-289-10

b. Send or Dust. Clean out the engine compartment (see paragraph 2-6) and have organiztional maintenance schedule a complete oil change and lubrication. Make sure maintenance knows it's due to operation in ssnd or dust. Check your air filter daily for dust and dirt. Have air filter replaced if it's dusty, dirty, or sandy. If the engine starts overheating, thoroughly clean radiator cooling fins. Also, be on the lookout for leaks in radiator.

3-11. AFTER FORDING

a. Fresh Water Fording. No special maintenance is required after fording fresh water 16 inches (41 cm) deep or less, but look for any deposits of sand or mud. If you find any sand or mud, clean it off. If the water YOU crossed was deeper than 20 inches (51 cm), notify your supervisor to schedule the truck for a complete oil change and lubrication. Make sure maintenance knows it's an after fording special maintenance.

b. Salt Water Fording. Salt causes almost immediate corrosion of metal surfaces. Clean entire truck (including underneath) with fresh water as soon as possible. If the salt water was mora than 20 inches (51 cm) deep, notify your supervisor to schedule an immediate complete oil change and lubrication. Make sure maintenance knows it's an after fording salt water maintenance.

TM 9-2320-289-10

APPENDIX A
REFERENCES

A-1. SCOPE

This appendix lists all forms, field manuals, technical manuals, and misc. pubs. referenced in this manual.

A-2. PUBLICATION INDEXES

The following indexes should be consulted frequently for the latest changes or revisions and for new publications relating to materiel covered in this technical manual.

Consolidated Index of Army Publications and Blank Forms . . . . DA Pam 25-30
US Army Equipment Index of Modification Work Orders . . . . DA Pam 750-10

A-3. FORMS

Refer to DA Pam 310-1 for a current and complete list of blank forms. DA Pam 738-750, The Army Maintenance Management Systems (TAMMS), contains instructions on the use of maintenance forms pertaining to this materiel.

A-4. OTHER PUBLICATIONS

    a. Decontamination.

Nuclear, Biological, and Chemical (NBC) Decontamination . . . . FM 3-5

Operator's and Organizational Maintenance Manual (including
  Repair Parts and Special Tools Lists): for Decontaminating
  Apparatus, Portable, DS2, 1-1/2 quart, ABC-M 11
  (NSN 4230-00-720-1618) . . . . TM 3-4230-204-12&P

    b. General.

Evacuation of the Sick and Wounded . . . . FM 8-35
First Aid for Soldiers . . . .
Manual for the Wheeled Vehicle Driver . . . . FM 21-11
Basic Cold Weather Manual . . . . FM 21-305
NorthernO aerations . . . . FM 31-70
Operation and Maintenance of Ordnance Materiel in
  Cold Weather (0°to-65°F) . . . . FM 31-71
Driver Selection and Training (Wheeled Vehicles) . . . . FM 9-207
    FM 55-30
Operator's, Organizational, Direct Support and General Support Main-tenance
  Manual (including Repair Parts and Special Tools Lists) for Trailer, Cargo: 3/4
  Ton, 2 Wheel, M101(2330-00-738-9509),
  M101A1 (2330-00-898-6779), M102A2 (2330-01-102-4697), and
  Chassis: Trailer: 3/4 Ton, 2 Wheel. M116(2330-00-898-6780)
  and M116A2 (2330-01-101-8434) . . . . TM 9-2330-202-14&P
Operator's and Organizational Maintenance Manual for
  Position and Azimuth Determining System AN/USQ-70
  Part Number 880500-1 (NSN 6675-01-071-5552) . . . . TM 5-6675-308-12

    c. Maintenance.

Lubrication Order for Truck, Cargo, Tactical, 4X4, M 1008,
  M1008A1, M1009, MI0I0, M1028, M1028A1, and M1031 . . . . LO 9-2320-289-12 Procedures for Destruction of Tank-Automotive Equipment
  to Prevent Enemy Use . . . . TM750-244-6 Use of Anti-freeze Solutions and Cleaning Compounds in
  Engine Cooling Systems . . . . TB 750-651

Change 3 A-1/(A-2   blank)

TM 9-2320-289-10

APPENDIX B
COMPONENTS OF END ITEM
AND BASIC ISSUE ITEMS LISTS

Section I. INTRODUCTION

B-1. SCOPE

This appendix lists Components of End Item and Basic Issue Items for the CUCV Series truck to help you inventory items required for safe and efficient operation.

B-2. GENERAL

NOTE

CUCV Series trucks do not have Components of End Item currently assigned.

The Basic Issue Items (BII) List is found in Section II and identifies the minimum essential items required to place the truck in operation, to operate it, and to perform emergency repairs. Although shipped separately, packaged BII must be with the truck during operation and whenever it is transferred between property accounts. The illustrations will assist you with hard-to-identify items. This appendix is your authority to request/requisition replacement BII, based on TOE/MTOE authorization of the end item.

B-3. EXPLANATION OF COLUMNS

Below is an explanation of columns found in the tabular listings:

    a. Column (1) - Illustration Number (Illus Number). This column indicates the number of the illustration which shows the item.

    b. Column (2) - National Stock Number. Indicates the National Stock Number assigned to the item and will be used for requisitioning purposes.

    c. Column (3) - Description. Indicates the Federal item name and, if required, a minimum description in parentheses to identify and locate the item. The entry for each item ends with the Federal Supply Code for Manufacturer (FSCM) in parentheses followed by the part number. Used On Code indicates the truck to which the item is assigned. For an explanation of these codes, refer to paragraph C-3.

    d. Column (4) - Unit of Measure (U/M). Indicates the measure used in performing the actual operational/maintenance function. This measure is expressed by a two-character alphabetical abbreviation (e.g., ea, in, pr).

    e. Column (5)- Quantity Required (Qty Rqr). Indicates the quantity of the item authorized to be used with the equipment.

TM 9-2320-289-10

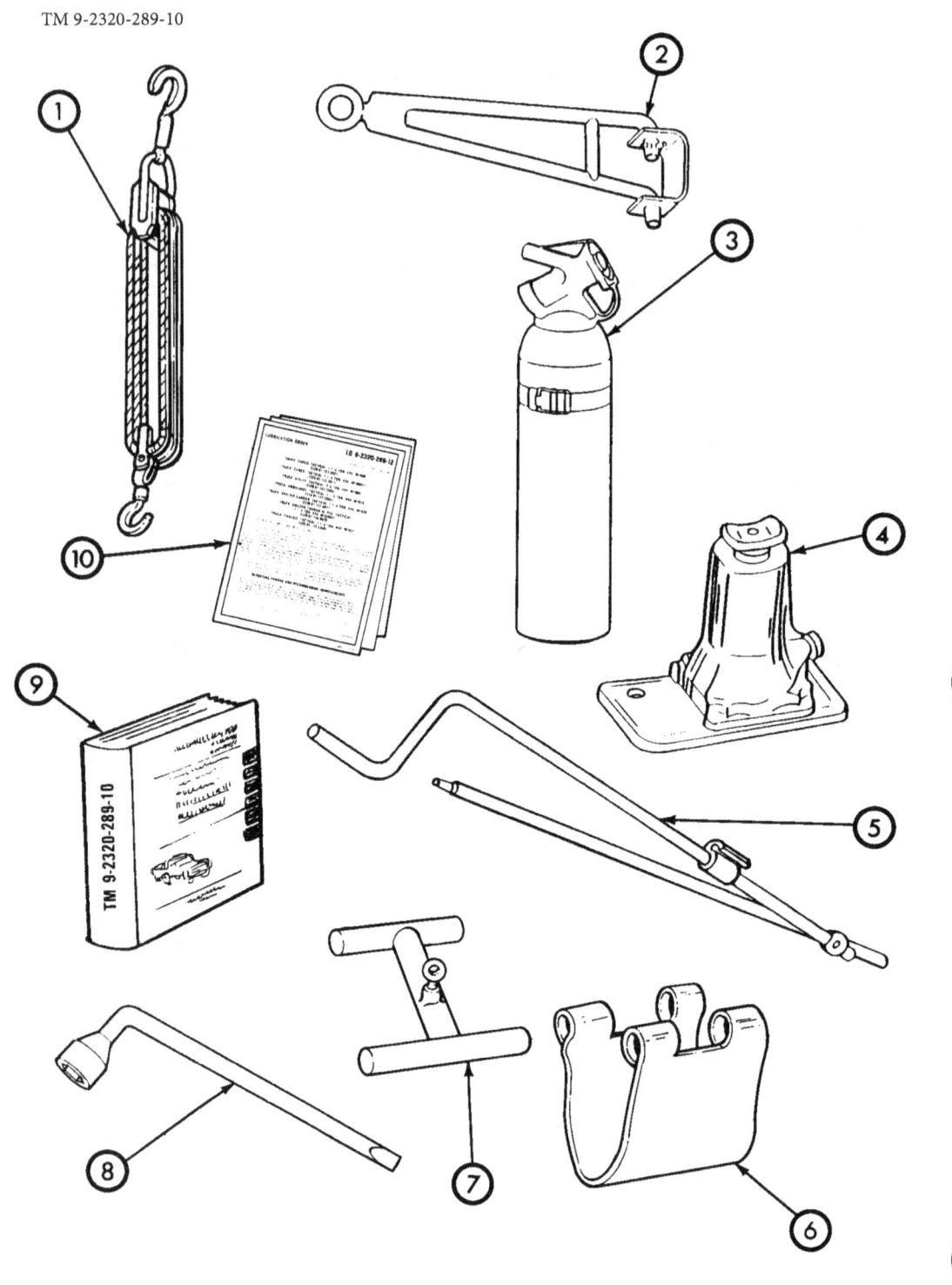

Figure B ll.

Section II. BASIC ISSUE ITEMS LIST

(1) Illus Number	(2) National Stock Number	(3) Description, FSCM & Part Number	Used On Code	(4) U/M	(5) Qty Rqr
1	2510-01-165-4540	Block and Tackle: for Patient Assist (25022) 22-0882	210	ea	1
2	2540-01-164-6171	Boom, Patient Assist (25022) 99-4316-0	210	ea	1
3	4210-00-889-2221	Extinguisher, Fire, Monobro, Hand: (16236) CS 4210-0009 CEFN	210	ea	1
4	5120-01-160-8868	Jack (11862) 14018603	ALL	ea	1
5	5120-01-164-0173	Jack Handle (11862) 14007107	ALL	ea	1
6	2510-01-165-4539	Sling, Litter: for Patient Assist (25022) 22-0870	210	ea	1
7	2590-01-200-5808	Sling, Spreader (25022) 99-4392-0	210	ea	1
8	5120-01-163-4726	Wheel Wrench (11862) 14036400	209	ea	1
		OR			
	5120-01-156-7296	Wheel Wrench (11862) 14009303	ALL EXC 209, 254, and 256	ea	1
		OR			
		Wheel Wrench (11862) 14064610	254 and 256	ea	1
9		TM 9-2320-289-10 Manual: Technical, Operator's (in instrument panel compartment)	ALL	ea	1
10		LO 9-2320-289-12 Order: Lubrication (in instrument panel compartment)	ALL	ea	1

TM 9-2320-289-10

APPENDIX C
ADDITIONAL AUTHORIZATION LIST

Section I. INTRODUCTION

C-1. SCOPE

This appendix lists additional Items that you are authorized for the support of the Truck, Car-go, Tactical, 4X4, M1008, M1008A1, M1009, M1010, M1028, M1028A1, M1028A2, M1028A3, and M1031 trucks.

C-2. GENERAL

This list identifies items thst do not hsve to accompany the truck and that do not have to be turned in with it. These items are authorized to you by CTA, MTOE, TDA, or JTA.

C-3. EXPLANATION OF LISTING

National Stock Numbers, description, and quantities are provided to help you identify and request the additional items you requira to support this equipment. If the item required differs for different models of this equipment, see the "Used On Code" column for the applicable model or models. Codes used are:

USED ON CODE	MODEL
ALL 194	ALL
208	M 1008
209	M1008A1
210	M1009
230	M1010
231	M1028
252	M1031
254	M1028A1
256	M1028A2
	M1028A3

Section II. ADDITIONAL AUTHORIZATION LIST

(1) National Stock Number	Description FSCM & Part Number	Used On Code	(3) U/M	(4) Qty Auth
5935-00-322-8959	Adapter, Connector (19027) 11677570	ALL	ea	2
4730-00-808-5089	Adapter, Pipe Straight (81361) B5-19-1676-2	210		23
4730-00-808-5090	Adapter, Pipe Straight (81361) B5-19-1676-1			2
6665-00-935-6955	Alarm, Agent (81361) C5-15-8803	ALL		1
5110-00-293-2336	Ax: Single Bit, 4 lb. Head Weight, 36-1/2" Long, Type 1, Class 1, Design A (19207) 6150925	209		
2540-00-670-2459	Bag Assembly, Pamphlet Cotton Duck, 3" x 9-1/4" x 1-1/4" (19207) 11676920	ALL		
5140-00-473-6256	Bag, Tool, Cotton Duck, 10" x 20" with Flap (34623) 11655979			
2540-00-791-3343	Block, Chock (19207) 10860840			2
5340-00-595-5208	Bracket, Angle (81361) B5-19-1831	210		7
2590-00-148-7961	Cable, NATO Slave, 20' with Adapter, Connector, 2 Prong (19207) 11682379-1	ALL		1
6150-01-022-6004	Cable, NATO Slave with End Connectors 11682336-1			
2590-00-398-6527	Cable, 20' End Connectors 11682337-1			
2540-01-185-8306	Chain, Tire, 15" (16457) 3210	209	se	
2540-00-528-7360	Chain, Tire, 16" (72671) V2812	ALL EXC 209		
4730-00-554-7208	Clamp, Hose (98441) 10181-20	210	ea	16
	Clamp, Hose (96906) MS 22064-5			30
1025-00-007-9453	Connector, Orifice (81361) B5-19-1829			7
2510-00-567-0128	Connector, Slave Cable End (19207) 11682338	ALL		2

TM 9-2320-289-10

Section II. ADDITIONAL AUTHORIZATION LIST - Continued

(1) National Stock Number	(2) Description FSCM & Part Number	Used On Code	(3) U/M	(4) Qty Auth
4730-00-935-1643	Coupling, Half Quick (81361) C5-19-1900	210	ea	7
4230-00-720-1618	Apparatus (81361) Decontamination ABC M-11	ALL	ea	1
4230-01-133-4124	D5-51- Apparatus (81361) E5-51-527 M-13	ALL	ea	1
4210-00-889-2221	Extinguisher, Fire, Monobro, Hand (16236) CS 4210-0009 CEFN	ALL EXC 210	ea	1
4240-00-828-3952	Filter, D5-19-2350	210	ea	2
4240-01-026-3112	Filter, Assy. D5-19-2353	210	ea	1
4240-00-565-6059	Assy. D5-19-1754	210	ea	1
4910-00-204-3170	Tire 7188BH	ALL	ea	1
5120-00-288-6574	Handle, Mattock Pick, 36" Long (19207) 11677021	209	ea	1
4240-00-807-6856	Heater, Air Electric (81361) D5-19-1782	210	ea	5
4720-00-829-2761	Hose, Air Duct (81361) C5-19-916-4	210	ea	8
4720-00-279-0093	Hose, Air Duct (81361) C5-19-916-2	210	ea	1
4720-00-279-0094	Hose, Air Duct (81361) C5-19-916-3	210	ea	1
4720-00-829-2760	Hose, Air Duct (81361) C5-19-916	210	ea	5
4720-00-541-1313	Hose, Nonmetallic (98441) 111-20	210	ft	1
	Hose, Rubber, 2" Long M/F 4720-00-541-1313	210	ea	6
	Hose, Rubber, 6-1/2" Long M/F 4720-00-541-1313	210	ea	1
	Hose, Rubber, 11" Long M/F 4720-00-541-1313	210	ea	1
6545-00-922-1200	Kit, First Aid: 2-1/2" x 6-3/4" x 9-1/2" (19207) 11677011	ALL	ea	1

C-3

Section II. ADDITIONAL AUTHORIZATION LIST - Continued

(1) National Stock Number	(2) Description FSCM & Part Number	Used On Code	(3) U/M	(4) Qty Auth
9905-00-148-9546	Kit, Warning Device (81348) RR-W-1817	ALL	ea	1
6530-00-783-7905	Litter, (90142) 51 L 1976	210	ea	4
5120-00-243-2395	Mattock: Pick, Head, 19-1/2" Long (19207) 11677022	209	ea	1
5120-00-223-7397	Pliers, Slip Joint with Cutter, 8" (19207) 5214421	ALL	ea	1
5365-00-514-0393	Ring, Retaining (96906) MS 16624-4087	210	ea	7
5120-00-240-8716	Screwdriver, Crosstip, 3" (55719) SSDP31, 6	ALL	ea	1
5120-00-596-8502	Screwdriver, Flat 1-1/2" GGG-S-121	ALL	ea	1
5120-00-237-6985	Flat Tip, 8" 10510988	ALL	ea	1
4910-00-437-7215	Tire Bead, 15" 93-240	209	ea	1
4910-01-022-9721	Tire 8ead, 16-1/2" (01717) Model TC-50	ALL EXC 209	ea	1
5120-00-293-3336	Shovel: Hand, Round Point, 40-1/2" Long (19207) 11655784	209	ea	1
5120-00-240-5328	Wrench, Adjustable, Open End, 8" Long (92878) 1500559	ALL	ea	1

APPENDIX D
EXPENDABLE/DURABLE SUPPLIES
AND MATERIALS LIST

Section I. INTRODUCTION

D-1. SCOPE

This appandix lists expendable/durable supplies and materials that you will need to operate and maintain the truck. These items are authorized to you by CTA 50-970, Expendable Items.

D-2. EXPLANATION OF COLUMNS

a. Column (1) - Item Number. This number is assigned to the entry in the listing.

b. Column (2) - Level. This column identifies the lowest level of maintenance that requires the listed item.

    C - Operator/Crew

c. Column (3) - National Stock Number. This is the National Stock Number assigned to the item; use it to request or requisition the item.

d. Column (4) - Description. Indicates the Federal item name and, if required, a description to identify the item. The last line for each item indicates the Federal Supply Code for Manufacturer (FSCM) in parentheses, followed by the part number, if applicable.

e. Column (5) - Unit of Measure (U/M). Indicates the measure used in performing the actual maintenance function. This measure is expressed by a two-character alphabetical abbreviation (e.g., ea, in, pr). If the unit of measure differs from the unit of issue, requisition the lowest unit of issue that will satisfy your requirements.

TM 9-2320-289-10

Section II. EXPENDABLE/DURABLE SUPPLIES AND REQUIREMENTS LIST

(1) Item Number	(2) Level	(3) National Stock Number	(4) Description	(5) U/M
1.	C	6850-00-174-1806	Antifreeze: Arctic 55 Gallon Drum (81349) MlL-A-l 1755	gal
2.	C	6850-00-181-7929	Antifreeze: Ethylene Glycol, Inhibited, Heavy Duty, Single Package, 1 Gallon Can (81349) MIL-A-46153	gal
3.	C	6850-00-181-7933	Antifreeze: Ethylene Glycol, Inhibited, Heavy Duty, Single Package, 5 Gallon Can (81349)	gal
4.	C	6850-00-181-7940	Antifreeze: Glycol, Inhibited, Heavy Duty, Single Package, 55 Gallon Drum (81349) MIL-A-46153	gal
5.	C	9150-01-102-9455	Brake Fluid: Silicone, Automotive, All Weather Operational and Preservative, 1 Gallon Can (81349) MIL-B-46176	gal
6.	C	9150-01-123-3152	Brake Fluid: Silicone, Automotive, All Weather Operational and Preservative, 5 Gallon Can (81349) MIL-B-46176	gal
7.	C	9150-01-072-8379	Brake Fluid: Silicone, Automotive, 55 Gallon Drum (81349) MIL-B-46176	gal
8.	C	6850-00-926-2275	Cleaning Compound: Wind- Washer, 1 Pint Bottle (81348) O-C- 901	pt
9.	C	7930-00-282-9699	Detergent: Purpose, Liquid, 1 Gallon Can (81349) MIL-D-16791	gal
10.	C	6850-00-110-4498	Dry Cleaning Solvent: 1 Pint Can (81348) P-D-680	pt
11.	C	6850-00-274-5421	Dry Cleaning Solvent: 5 Gallon Can (81348) P-D-680	gal
12.	C	6850-00-285-8011	Dry Cleaning Solvent: 55 Gallon Drum (81348) P-D-680	gal

Section II. EXPENDABLE/DURABLE SUPPLIES AND REQUIREMENTS LIST - Continued

(1) Item Number	(2) Level	(3) National Stock Number	(4) Description	(5) U/M
13.	c	9140-00-286-5295	Fuel Oil, Diesel: Regular, DF-2, 5 Gallon Can (81348) VV-F-800	gal
14.	c	9140-00-286-5296	Fuel Oil, Diesel: Regular, DF-2, 55 Gallon Drum (81348) VV-F-800	gal
15.	c	9140-00-286-5287	Fuel Oil, Diesel: Winter, DF-1, 5 Gallon Can (81348) VV-F-800	gal
16.	c	9140-00-286-5288	Fuel Oil, Diesel: Winter, DF-1, 55 Gallon Drum VV-F-800	gal
17.	c	9140-00-286-5282	Fuel Oil, Diesel: Arctic, DF-A, 5 Gallon Can (81348) VV-F-800	gal
18.	c	9140-00-286-5284	Fuel Oil, Diesel: Arctic, DF-A, 55 Gallon Drum VV-F-800	gal
19.	c	9150-00-935-1017	Grease: Automotive and Artillery, 14 Ounce Cartridge (81349)	oz
20.	c	9150-00-190-0904	Grease: and Artillery, 1-3/4 Pound Can (81349)	lb
21.	c	9150-00-190-0905	Grease: and Artillery, 6-1/2 Pound Can (81349) MIL-G-10924	lb
22.	c	9150-00-698-2382	Hydraulic Fluid: Transmission, 1 Quart Can (24617) Dexron® II	qt
23.	c	9150-00-657-4959	Hydraulic Fluid: Transmission, 5 Gallon Can (24617) Dexron® II	gal
24.	c	9150-01-035-5390	Lubricating Oil: Gear, Multipurpose, GO 75W, 1 Quart Can (81349) MIL-L-2105	qt

TM 9-2320-289-10

Section II. EXPENDABLE/DURABLE SUPPLIES AND REQUIREMENTS LIST - Continued

(1) Item Number	(2) Level	(3) National Stock Number	(4) Description	(5) U/M
25.	c	9150-01-035-5391	Lubricating Oil: Gear, Multipurpose, GO 75W, 5 Gallon Can (81349) MIL-L-2105	gal
26.	c	9150-01-035-5392	Lubricating Oil: Gear, Multipurpose, GO 80 W/90, 1 Quart Can (81349) MIL-L-2105	qt
27.	c	9150-00-001-9395	Lubricating Oil: Gear, Multipurpose, GO 80 W/90, 5 Gallon Can (81349) MIL-L-2105	gal
28.	c	9150-00-231-6689	Lubricating Oil: General Purpose, Preservative, PL-S, 1 Quart Can (81348) VVL800	qt
29.	c	9150-00-402-4478	Lubricating Oil: Internal Combustion Engine, Arctic, OEA, 1 Quart Can (81349) MIL-L-46167	qt
30.	c	9150-00-402-2372	Lubricating Oil: Internal Combustion Engine, Arctic, OEA, 5 Gallon Can (81349)	gal
31.	c	9150-00-491-7197	Oil: Internal Engine, Arctic, OEA, 55 Gallon Drum (81349) MIL-L-46167	gal
32.	c	9150-00-189-6727	Lubricating Oil: Internal Combustion Engine, Tactical Service, OE/HDO 10W, 1 Quart Can (81349) MIL-L-2104	qt
33.	c	9150-00-186-6668	Lubricating Oil: Internal Combustion Engine, Tactical Service, OE/HDO 10W, 5 Gallon Can (81349) MIL-L-2104	gal
34.	c	9150-00-191-2772	Lubricating Oil: Internal Combustion Engine, Tactical Service, OE/HDO 10W, 55 Gallon Drum (81349) MIL-L-	gal

TM 9-2320-289-10

Section II. EXPEINABLE/DURABLE SUPPLIES AND REQUIREMENTS LIST - Continued

(1) Item Number	(2) Level	(3) National Stock Number	(4) Description	(5) U/M
35.	c	9150-00-188-6681	Lubricating Oil: Internal Combustion Engine, Tactical Service, OE/HDO 30W, 1 Quart Can (81349) MIL-L-2104	qt
36.	c	9150-00-188-9858	Lubricating Oil: Internal Combustion Engine, Tactical Service, OE/HDO 30W, 5 Gallon Can (81349) MIL-L-2104	gal
37.	c	9150-00-189-6729	Lubricating Oil: Internal Combustion Engine, Tactical Service, OE/ 30W, 55 Gallon Drum (81349) MIL-L-2104	gal
38.	c	9150-01-152-4117	Lubricating Oil: Internal Combustion Engine, Tactical Service, OE/HDO 15W/40, 1 Quart Can (81349) MIL-L-2104	qt
39.	c	9150-01-152-4118	Lubricating Oil: Internal Combustion Engine, Tactical Service, OE/HDO 15W/40, 5 Gallon Can (81349) MIL-L-2104	gal
40.	c	9150-01-152-4119	Lubricating Oil: Internal Combustion Engine, Tactical Service, OE/HDO 15W/40, 55 Gallon Drum (81349) MIL-L-2104	gal
41.	c	7920-00-205-1711	Rag: Wiping Cotton and Cotton-Synthetic (58536) A-A-53 1, 50 Pound Bale	lb
42.	c	6850-01-160-3868	Inhibitor, Corrosion, Cooling Liquid (81349) MIL-A-53009	qt

D-5/(D-6 blank)

TM 9-2320-289-10

APPENDIX E
STOWAGE AND SIGN GUIDE
FOR COMPONENTS OF END ITEM, BASIC ISSUE ITEMS,
AND APPLICABLE ADDITIONAL AUTHORIZATION LIST ITEMS

E-1. SCOPE

This appendix shows the location for stowage of equipment and material required to be carried on the CUCV Series trucks.

E-2. GENERAL

The pictures below and on the following pages show the location of decals, stencils, and metal signs used on the truck. Some are cautions or information that you need to safely operate the truck. Signs outlined with boxes are decals, signs not outlined are stencils, and signs outlined with dotted boxes are metal stamped.

NOTE

Decal locations are the sama on all trucks unless specifically noted on model variations.

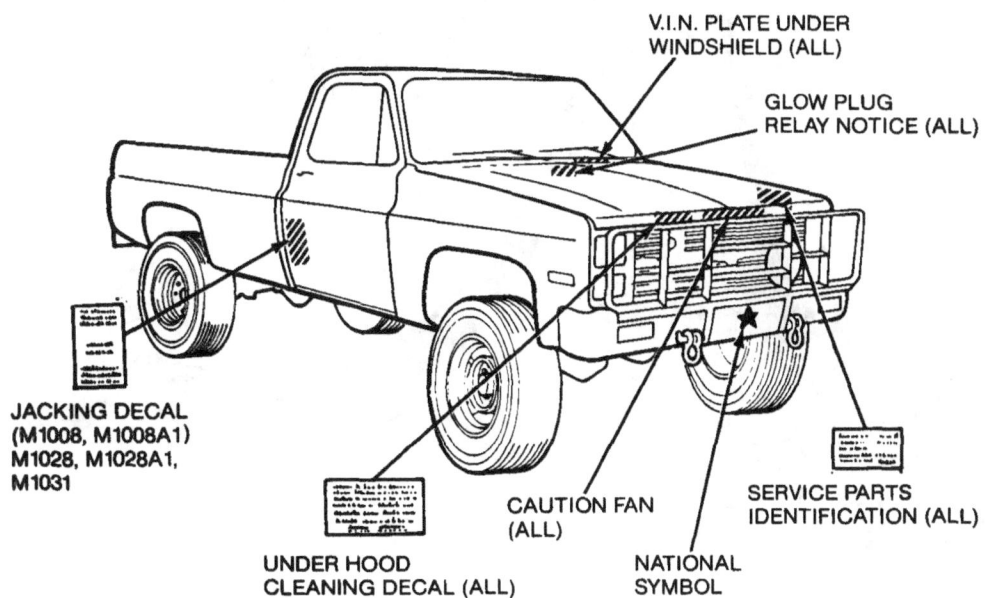

E-1

TM 9-2320-289-10

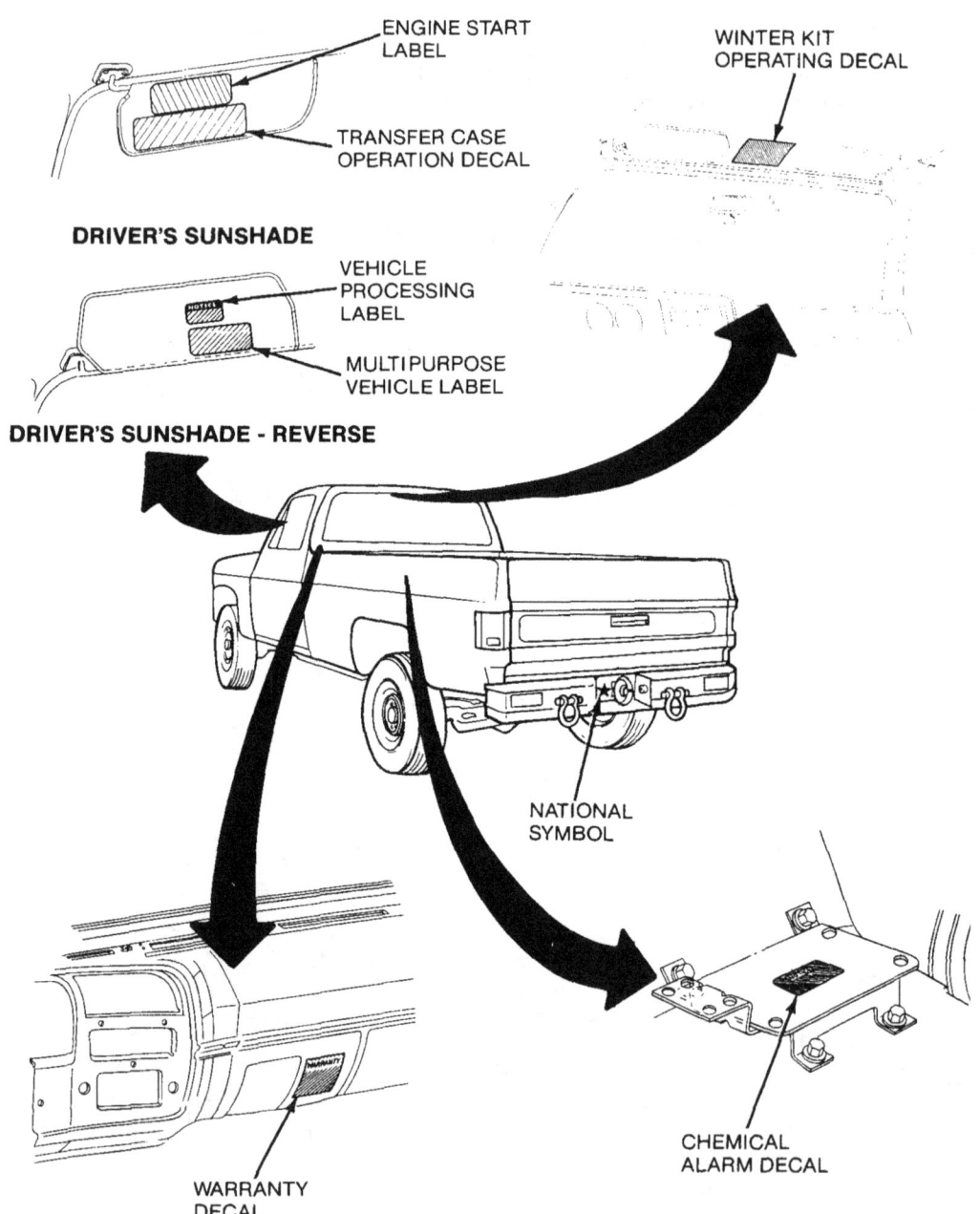

All Models

TM 9-2320-289-10

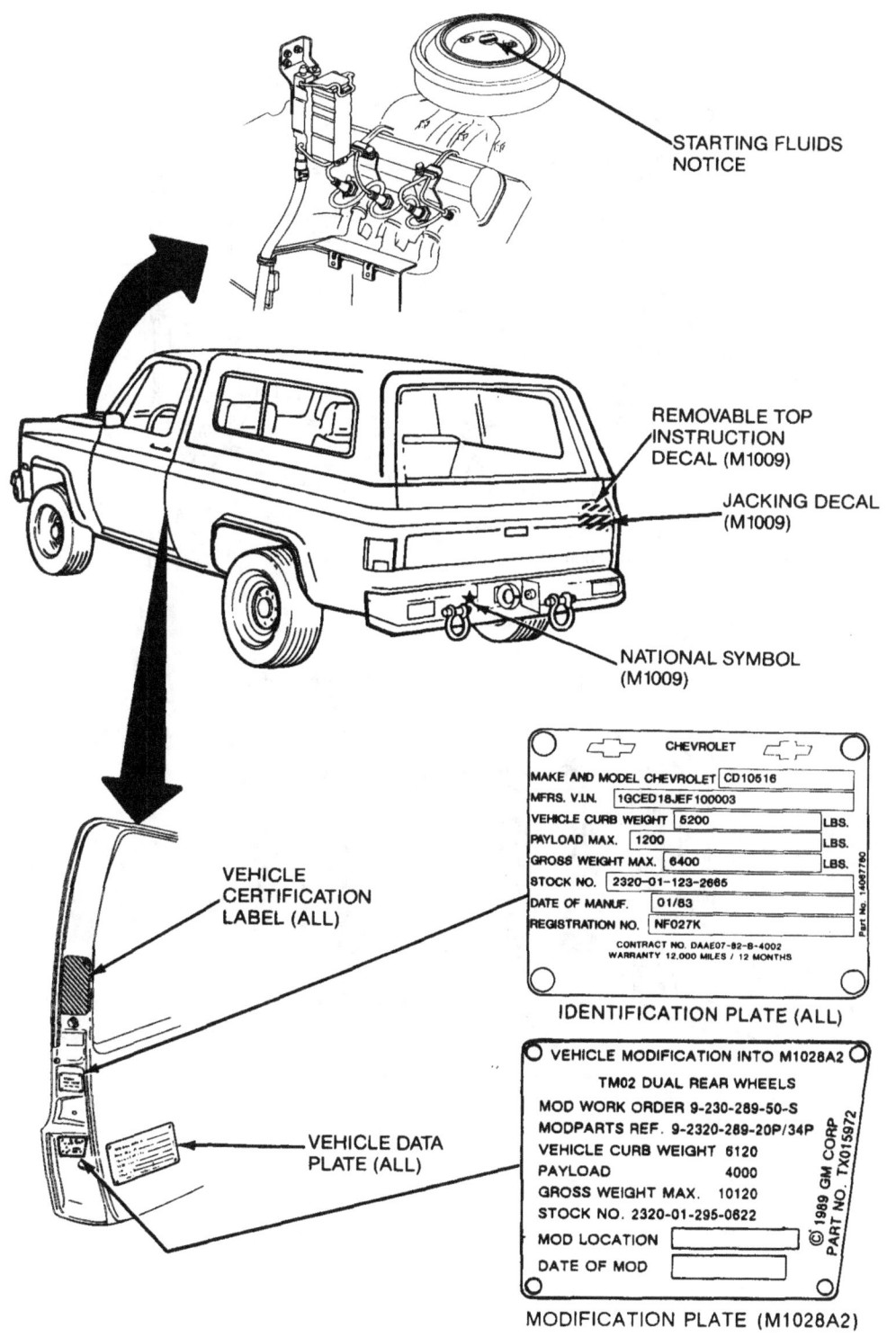

TM 9-2320-289-10

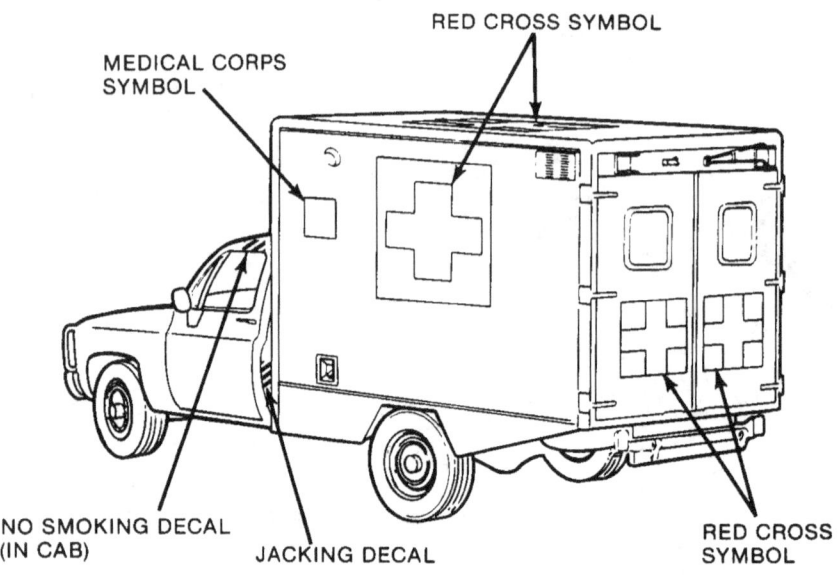

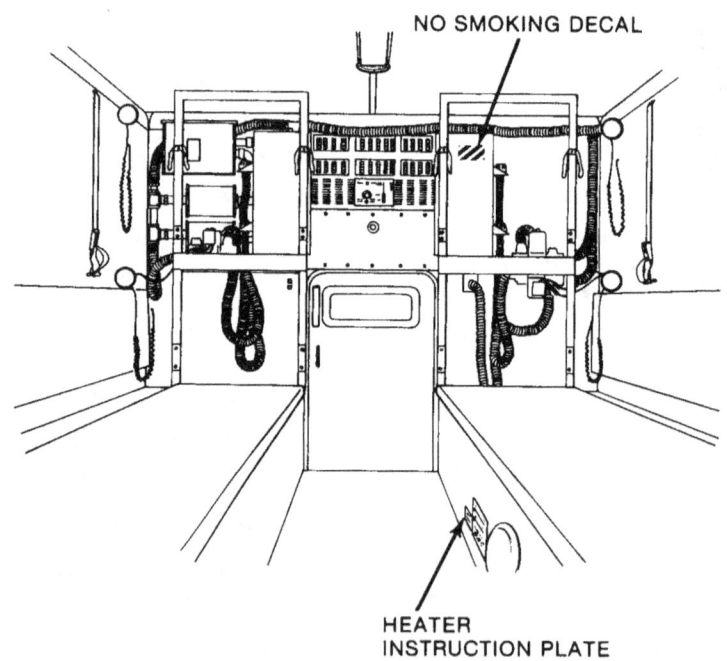

M1010 Only

TM 9-2320-289-10

INDEX

SUBJECT	Paragraph	Page

A

Subject	Paragraph	Page
Abbreviations	1-5	1-1
Accelerator pedal	2-2(16)	2-3
Access door, heater	1-7	1-7
Access steps:		
Location	1-7	1-7
Operation	2-30	2-77
Additional authorization list (AAL)	C-1	C-1
After fording maintenance	3-11	3-15
After operating on unusual terrain	3-10	3-14
Agents, cleaning	2-6	2-6
Air conditioner (M1010), operation	2-26	2-67
Aircraft, towing	2-14(b)	2-57
Air exhaust vent	1-7	1-7
Air inlet, heater	1-7	1-7
Air intake manifold	1-10(a)	1-13
Air vent, operation	2-22	2-62
Ambulance:		
General	1-6(f)	1-4
Operation of peculiar components:		
Access steps	2-30	2-77
Air conditioner	2-26	2-67
Attendant's seat	2-29	2-76
Blackout curtains	2-33	2-78
Domelight	2-32	2-78
Floodlights	2-35	2-79
Focus lights	2-31	2-77
Gas-particulate filter unit	2-25	2-65
Litter berth	2-28	2-70
Litter tie-down	2-28(c)	2-75
Heater, patient compartment	2-27	2-68
Spotlight	2-34	2-78
Peculiar components:		
Air conditioning unit	1-17(b)	1-19
Domelight and focus lights	1-17(d)	1-19
Gas-particulate filter unit	1-17(a)	1-18
Heater, personnel	1-17(c)	1-19
Attendant's seat (M1010)	2-29	2-76

B

Subject	Paragraph	Page
Basic issue items (BII)	B-1	B-1
Battery system	1-14(a)	1-16
Berth, litter (M1010), operation	2-28	2-70
Blackout		
Curtains, operation	2-33	2-78
Drive lights, operation	2-15(b)	2-58
Markers, operation	2-15(c)	2-58
Block and tackle (M1010)	2-28	2-73
Boom, patient assist	1-7	1-7
Bracket, patient assist boom	1-7, 2-28	1-7, 2-72
Brake:		
Parking, operation	2-13(b)	2-57
Pedal	2-2(26)	2-3
Release	2-9	2-54
Release handle	2-2(23)	2-3

Index 1

TM 9-2320-289-10

INDEX

SUBJECT	Paragraph	Page
Service, pedal:		
Location	2-2(25)	2-3
Operation	2-10	2-54
System	1-15	1-17
System warning light	2-2(17)	2-3

C

Cab controls, operation:		
Air vents		
Hazard warning flasher	2-22	2-62
	2-21	2-62
Headlight dimmer switch	2-16	2-58
Headlight/pa parking light switch	2-15	2-58
Heater/defroster controls	2-20	2-61
Ignition switch	2-8	2-53
Parking brake pedal	2-13(b)	2-57
Service brake pedal	2-10	2-54
Transfer case control lever	2-12	2-55
Windshield wiper/washer control	2-18	2-60
Capabilities, equipment	1-6	1-3
Capacities, tabulated data	1-9	1-12
Cargo box cover kit	2-37	2-83
Cargo, truck	1-6(d)	1-3
Changing tires	3-6	3-9
Characteristics, equipment	1-6	1-3
Charging system	1-14(a)	1-16
Chassis, truck	1-6(g)	1-5
Classification marker, weight	1-7	1-6
Cleaning agents:		
Cleaning rust or grease	2-6(c)	2-7
Cleaning underwood areas	2-6(a)	2-6
Treating mildewed areas	2-6(b)	2-7
Cold weather operations:		
Abnormal instrument readings	2-42	2-89
Halt or parking	2-42	2-89
Long halts or extended shutdown	3-8(c)	3-14
Maintenance	3-8	3-14
Preparing truck	3-8(a)	3-14
Snow and ice and mud	2-42	2-89
Startup	2-42	2-88
Columns, components of end item and BII, explanation of	B-3	B-1
Columns, expendable/durable supplies and materials list, explanation of	D-2	D-1
Controls and indicators, description and use of:		
Accelerator pedal	2-2(16)	2-3
Blackout drive switch	2-2(24)	2-3
Brake release handle	2-2(23)	2-3
Brake system warning light	2-2(17)	2-3
Door ajar indicator light (M1010 only)	2-2(11)	2-2
Engine coolant temperature light	2-2(4)	2-2
Floodlight controls (M1010 only)	2-2(15)	2-3
Four-wheel drive indicator light	2-2(2)	2-2
Fuel gage	2-2(10)	2-3
Gas-particulate filter unit controls (M1010 only)	2-2(14)	2-2
Gen 1 and Gen 2 lights	2-2(1)	2-3
Hazard warning flasher	2-2(19)	2-3
Headlight high beam indicator light	2-2(6)	
Heater/defroster controls	2-2(13)	
	2-2(27)	

Index 2

TM 9-2320-289-10

INDEX

SUBJECT	Paragraph	Page
Light switch	2-2(21)	2-3
Low coolant warning light	2-2(20)	2-3
Oil pressure light	2-2(3)	2-2
Parking brake pedal	2-2(26)	2-3
Seat belt indicator light	2-2(9)	2-2
Service brake pedal	2-2(25)	2-3
Service lights/blackout toggle switch	2-2(22)	2-3
Speedometer	2-2(5)	2-2
Turn signal indicators	2-2(18)	2-3
Voltmeter	2-2(12)	2-2
WAIT (glow plugs indicator) light	2-2(7)	2-2
WATER-IN-FUEL indicator light	2-2(8)	2-2
Controls, instruments, gages, and lights	2-2	2-1
Coupling, trailer electrical	1-7	1-6
Cover kit, cargo box	2-37	2-83
Curtains, blackout, operation	2-33	2-78

D

Data, tabulated	1-9	1-11
Defroster, heater/operation	2-20	2-61
Description of equipment and major components	1-6, 1-7	1-3, 1-6
Differences between models	1-8	1-11
Dimmer switch, headlight	2-16	2-58
Disabled truck, tow	2-41	2-87
Distribution, fuel	1-10(a)	1-13
Distribution, load	1-18	1-19
Domelight operation	2-32	2-78
Drain fuel filter	3-7	3-12
Drive truck:		
Forward	2-9	2-54
In four-wheel drive	2-12	2-54
In reverse	2-11	2-54
Driving tips:		
Transfer case	1-13	1-16
Transmission	1-12	1-14
Dual wheel configuration	1-20	1-20

E

Electrical system:

Battery system	1-14(a)	1-16
Charging system	1-14(a)	1-16
Glow plug system	1-14(a)	1-17
Starting system	1-14(a)	1-16
Wiring and lighting system	1-14(a)	1-17
Emergency procedures:		
Slave starting	2-40	2-86
Towing	2-41	2-87
Engine:		
Air intake manifold	1-10(a)	1-13
Cold weather starting	2-42	2-88
Fuel distribution	1-10(a)	1-13
General	1-10	1-13
Hot weather starting	2-43	2-89
Ignition	1-10(a)	1-13
Normal starting	2-8	2-53

Change 3   Index 3

TM 9-2320-289-10

INDEX

SUBJECT	Paragraph	Page
Shut down	2-13	2-56
Slave starting	2-40	2-86
Troubleshooting symptom index	3-5	3-3
Equipment characteristics and capabilities:		
Ambulance	1-6(f)	1-4
Cargo truck	1-6(d)	1-3
Chassis truck	1-6(g)	1-5
Utility truck	1-6(e)	1-4
Equipment improvement recommendations (EIR), reporting	1-3	1-1
Erect cargo box cover	2-37(b)	2-83
Expendable supplies and materials list	D-1	D-1
Explanation of columns, components of end item and BII:		
Column (1)-illustration number	B-3	B-1
Column (2)-national stock number	B-3	B-1
Column (3)-description	B-3	B-1
Column (4)-unit of measure	B-3	B-1
Column (5)-quantity required	B-3	B-1
Explanation of columns, expendable/durable supplies and materials list		
Column (l)-item number	D-2	D-1
Column (2)-level	D-2	D-1
Column (3)-national stock number	D-2	D-1
Column (4)-description	D-2	D-1
Column (5)-unit of measure	D-2	D-1
Explanation of listing, additional authorization list	C-3	C-1
External components:		
Access steps	1-7	1-7
Air conditioner	1-7	1-7
Air exhaust vent	1-7	1-6
Blackout drive light	1-7	1-6
Brush guard	1-7	1-8
Dual gearwheel	1-7	1-8
Flared rear fender	1-7	1-7
Floodlights	1-7	1-7
Fuel filler cap	1-7	1-7
Heater access door	1-7	1-7
Heater air inlet	1-7	1-6
Locking hubs	1-7	1-7
Medical corps symbol	1-7	1-8
Offset front wheel hub	1-7	1-7
Patient assist boom	1-7	1-7
Patient assist boom bracket	1-7	1-8
Rear fender side marker	1-7	1-7
Red cross symbol	1-7	1-6
Service light	1-7	1-6
Side markers	1-7	1-7
Slave receptacle	1-7	1-6
Spotlight	1-7	1-6
Stoplights	1-7	1-6
Tailgate	1-7	1-8
Tailgate latch release	1-7	1-8
Tailgate window	1-7	1-8
Tailgate window crank	1-7	1-6
Tow hooks	1-7	1-6
Tow pintle	1-7	1-6
Trailer electrical coupling	1-7	1-6
Weight classification marker		

F

Features, equipment	1-6	1-3
Filter unit, gas-particulate, operation	2-25	2-65

Index 4 Change 3

TM 9-2320-289-10

INDEX

SUBJECT	Paragraph	Page
Flasher, hazard warning, operation	2-21	2-62
Floodlight, operation	2-35	2-79
Focus lights, operation	2-31	2-77
Fording:		
After fording maintenance:		
Fresh water	3-11(a)	3-15
Sal twater	3-11(b)	3-15
Depth	2-45	2-91
General	2-45	2-91
Forms	A-3	A-1
Forms and records, maintenance	1-2	1-1
Forward, drive truck	2-9	2-54
Fuel:		
Distribution	1-10(a)	1-13
Filter, draining	3-7	3-12
System	1-11	1-13

G

Gages	2-2	2-1
Gas-particulate filter:		
Air outlet	1-7	1-10
Canisters	1-7	1-10
Unit		
Heater	1-7, 2-25	1-10, 2-67
Location	1-7	1-10
Operation	2-25	2-65
General:		
AAL	C-2	C-1
BII	B-2	B-1
Lubrication instructions	3-1	3-1
PMCS	2-4	2-5
Stowage and sign guide	E-2	E-1
Generator indicator lights	2-2(1)	2-2
Glow plug system	1-14(a)	1-17
Gross axle weight rating (GAWR)	1-18	1-19
Gross vehicle weight rating (GVWR)	1-18	1-19

H

Handle, hood	2-2(27)	2-3
Hazard warning flasher, operation	2-21	2-62
Headlight dimmer switch, operation	2-16	2-58
Headlight/parkinglight switch	2-15	2-58
Heater		
Patient compartment (M1010), operation	2-27	2-68
Personnel (M1010), operation	1-17(c)	1-19
Swingfire:		
Aid truck start-up	2-47	2-94
Operation	2-46	2-92
Winterization kit, operation:		
Defroster/interiorheater	2-36(b)	2-81
Enclosed cargo area personnel heater	2-36(c)	2-82
Underwood heaters	2-36(a)	2-80
Heater/defroster, operation:		
Fan	2-20(c)	2-61

Index 5

TM 9-2320-289-10

INDEX

SUBJECT	Paragraph	Page
In snow and ice	2-20(d)	2-62
Lower lever	2-20(b)	2-61
Upper lever	2-20(a)	2-61
Hood handle	2-2(27)	2-3
Horn, operation of	2-19	2-60
Hot weather operations:		
At halt or parking	2-43	2-90
Humid weather	3-9(b)	3.14
Maintenance	3-9	3-14
Hubs, locking	1-7, 2-12	1-6, 2-54

I

Index, troubleshooting symptom	3-5	3-3
Indexes, publications	A-2	A-1
Indicators and controls	2-2	2-1
Indicator, turn signal, operation	2-17	2-59
Information, warranty	1-4	1-1
Inlet, heater, air	1-7	1-7
Install troop seat kit	2-38(a)	2-84
Instruction plates and decals	E-2	E-1
Instrument panel:		2-3
Accelerator pedal	2-2(16)	2-3
Blackout drive switch	2-2(24)	2-3
Brake release handle	2-2(23)	2-3
Brake system warning light	2-2(17)	2-1
Controls and indicators	2-2	2-2
Door ajar indicator light (M1010 only)	2-2(11)	2-2
Engine coolant temperature light	2-2(4)	2-3
Floodlight controls (M1010 only)	2-2(15)	2-2
Four-wheel drive indicator light	2-2(2)	2-2
Fuel gage	2-2(10)	2-3
Gas-particulate filter unit controls (M1010 only)	2-2(14)	2-2
Genland Gen 2 lights	2-2(1)	2-3
Hazard warning flasher	2-2(19)	2-2
Headlight high beam indicator light	2-2(6)	2-3
Heater/defroster controls	2-2(13)	2-3
Hood handle	2-2(27)	2-3
Light switch	2-2(21)	2-3
Low coolant warning light	2-2(20)	2-2
Oil pressure light	2-2(3)	2-3
Parking brake pedal	2-2(26)	2-3
Seat belt indicator light	2-2(9)	2-2
Service brake pedal	2-2(25)	2-3
Service lights/blackout toggle switch	2-2(22)	2-2
Speedometer	2-2(5)	2-3
Turn signal indicators	2-2(18)	2-2
Voltmeter	2-2(12)	2-2
WAIT (glow plug indicator) light	2-2(7)	2-2
WATER-IN-FUEL indicator light	2-2(8)	
Internal components:		
Air conditioner controls	1-7	1-10
Air exhaust vents	1-7	1-9
Domelight	1-7	1-9
Focus lights	1-7	1-9
Gas-particulate filter:		
Air outlet	1-7	1-10

Index 6

TM 9-2320-289-10

INDEX

SUBJECT	Paragraph	Page
Canisters	1-7	1-10
Unit	1-7	1-10
Unit heater	1-7	1-10
Heater outlet	1-7	1-9
Lower litter berths	1-7	1-9
Patient compartment front door	1-7	1-9
Rear door hold open	1-7	1-9
Upper litter berths	1-7	1-9
Upperlitter berth support	1-7	1-10

K

	Paragraph	Page
Kit		
Cargo box cover	2-37	2-83
Troop seat		
Install	2-38(a)	2-84
Operate	2-39	2-85
Remove	2-38(b)	2-84
Winterization		
General	1-19	1-20
Heater operation	2-36	2-80

L

	Paragraph	Page
Leakage definitions	2-7	2-7
Lights:		
Blackout drive	2-15(b)	2-58
Blackout markers	2-15(c)	2-58
Brake system warning	2-2(17)	2-3
Dome and focus	1-17(d)	1-19
Dome, operation	2-32	2-78
Flood, operation	2-35	2-79
Focus, operation	2-31	2-77
Head/parking, operation	2-15(a)	2-58
Service	2-15(a)	2-58
Spot, operation	2-34	2-78
Listing, explanation of (AAL)	C-3	C-1
Litter berth (M1010), operation	2-28	2-70
Litter tie-down, operation	2-28(c)	2-75
Loading, truck	1-18	1-19
Location and description of major components:		
External	1-7(a)	1-6
Internal	1-7(b)	1-9
Locking hub, lock	2-12	2-54
Lubrication order	3-2	3-1

M

Maintenance:

	Paragraph	Page
Afterfording	3-11	3-15
After operating on unusual terrain:		
Mud	3-1O(a)	3-14
Sand or dust	3-10(b)	3-14
Cleaning agents and precautions:		
Mildewed areas	2-6(b)	2-7
Rust or grease	2-6(c)	2-7

Index 7

INDEX

SUBJECT	Paragraph	Page
Underhood areas	2-6(a)	2-6
Extreme cold weather	3-8	3-14
Extreme hot or humid weather	3-9	3-14
Forms and records, maintenance	1-2	1-1
Recording repairs	1-2	1-1
Major components, location and description of:		
External	1-7(a)	1-6
Interanl	1-7(b)	1-9
Manifold, air intake	1-10(a)	1-13
Mildew, cleaning	2-6(b)	2-7
Model differences	1-8	1-11
Mount, weapon, operation	2-24	2-64
M1009, 0perate tailgate	2-23	2-63
MIOIO, operate peculiar components:		
Air conditioner	2-26	2-67
Gas-particulate filter system	2-25	2-65
Litter berths	2-28	2-70
Patient compartment heater	2-27	2-68

O

Oil pressure light	2-2(3)	2-2
Operate:		
Access steps	2-30	2-77
Air conditioner (M1010)	2-26	2-67
Air vents	2-22	2-62
Attendant's seat	2-29	2-76
Blackout curtains	2-33	2-78
Domelight	2-32	2-78
Floodlights	2-35	2-79
Focus lights	2-31	2-77
Gas-particulate filter system (M1010)	2-25	2-65
Hazard warning flasher	2-21	2-62
Headlight dimmer switch	2-16	2-58
Heater/defroster	2-20	2-61
Horn	2-19	2-60
In extreme cold	2-42	2-88
In extreme heat	2-43	2-89
In unusual terrain	2-44	2-90
Lights	2-15	2-58
Litter berths (M1010)	2-28	2-70
On unusual terrain	2-44	2-90
Patient compartment heater (M1010)	2-27	2-68
Spotlight	2-34	2-78
Swingfire heater	2-46	2-92
Tailgate (M1009)	2-23	2-63
Troop seat kit	2-39	2-85
Turn signal indicator	2-17	2-59
Weapons mount	2-24	2-64
Windshieldwiper/washer		2-80
Winterization kit heaters	2-36	
Other publications:		A-1
Decontamination	A-4	A-1
General	A-4	A-1
Maintenance	A-4	

TM 9-2320-289-10

INDEX

SUBJECT	Paragraph	Page

P

Parking brake:
    Operation .... 2-13(b)   2-57
    Pedal .... 2-2(26)   2-3
    Release 2-9   2-54
    Release handle .... 2-2(23)   2-3
Parking truck:
    Cold weather .... 2-42(b)   2-89
    General 2-13   2-56
    Hot weather .... 2-43(d)   2-90
    With trailer .... 2-14(c)   2-57
Park truck and shutdown engine 2-13   2-56
Patient compartment (M1010)
    General 2-28   2-70
    Heater(M1010) 2-27   2-68
Peculiar components, M1010:
    Air conditioning unit .... 1-17(b)   1-19
    Domelight and focus lights .... 1-17(d)   1-19
    Gas-particulate filter unit (GPFU) .... 1-17(a)   1-18
    Personnel heater .... 1-17(c)   1-19
Pedal:
    Accelerator .... 2-2(16)   2-3
    Parking Brake .... 2-2(26)   2-3
    Service Brake .... 2-2(25)   2-3
Preparing truck for use, road test .... 2-3(a)   2-3
Preventive maintenance checks and services (PMCS)
    Cleaning agents 2-6   2-6
    General 2-4   2-5
    Leakage definitions 2-7   2-7
    Table   2-8
    Procedures 2-5   2-5
Publication indexes A-2   A-1
Publications, other A-4   A-1

R

Recording repairs 1-2
Records and forms, maintenance 1-2
References A-1   1-1
Remove troop seat kit .... 2-38(b)   1-1
Reporting equipment improvement recommendations (EIR) 1-3   A-1
                                                                                                                2-84
                                                                                                                1-1
Reverse - 2-11   2-54
Road test .... 2-3(a)   2-3

S

Scope:
    Appendix A A-1   A-1
    Appendix B B-1   B-1
    Appendix C C-1   C-1
    Appendix D D-1   D-1
    Appendix E E-1   E-1
    General information 1-1   1-1
    Troubleshooting 3-3   3-2
Seat, attendant's (M1010), operation 2-29   2-76

Index 9

TM 9-2320-289-10

INDEX

SUBJECT	Paragraph	Page
Service brake pedal:		
Location	2-2(25)	2-3
Operation	2-10	2-54
Service lights, operation,	2-15(a)	2-58
Shut down engine and park truck	2-13	2-56
Sign and stowage guide		E-1
Slave start truck	2-40	2-86
Speeds, transfer case	1-13	1-16
	2-34	2-78
Spotlight operation		2-53
Start engine	2-8	1-16
Starting system	1-14(a)	1-18
Steering system	1-16	
Steps, access:		1-7
Location		2-77
	1-7	2-54
Stop truck	2-10	E-1
Stowage and sign guide		2-83
Stow cargo box cover	2-37(a)	
Swingfire heater:		
Operation		2-94
Correct flooded condition	2-46(c)	2-92
Prepare heater	2-46(a)	2-92
Start heater	2-46(b)	2-94
Start truck with heater	2-47	2-58
Switch, headlight dimmer, operation	2-16	3-3
Symptom index, troubleshooting	3-5	
System:		1-16
		1-17
Battery	1-14(a)	1-16
Brake	1-15	1-17
Charging	1-14(a)	1-13
Electrical	1-14(b)	2-65
Fuel		1-17
	2-25	1-11
Glow plug	1-14(a)	1-16
Starting	1-14(a)	1-18
Steering	1-16	1-17
Wiring and lighting	1-14(a)	

T

SUBJECT	Paragraph	Page
Tabulated data	1-9	1-11
Tailgate:		
Latch release	1-7	
M1008	1-7	
M1009, operation	2-23	
Window (M1009)	1-7	
Window crank (M1009)	1-7	
Tips, driving:		
Transfer case	1-13	
Transmission	1-12	
Tires, changing	3-6	
Tow atrailer/aircraft	2-14	
Tow disabled truck:		
On front wheels	2-41(c)	
On gearwheels	2-41(b)	
With tow bar	2-41(a)	
Trailer, towing	2-14	

TM 9-2320-289-10

INDEX

SUBJECT	Paragraph	Page
Transfer case:		
Driving tips	1-13	1-16
General	1-13	1-15
Neutral (N)	1-13	1-15
Selecting a transfer range	1-13, 2-44	1-15, 2-90
Shifting	1-13, 2-12	1-15, 2-55
Transfer case control lever	1-13	1-15
Transfer range selection	1-13, 2-44	1-15, 2-90
Transmission driving tips	1-12	1-14
Transmission range selection:		
Drive	1-12	1-14
Neutral	1-12	1-14
Park	1-12	1-14
Reverse	1-12	1-14
"1"	1-12	1-14
"2"	1-12	1-14
Troop seat kit		
Install	2-38(a)	2-84
Operate	2-39	2-85
Remove	2-38(b)	2-84
Troubleshooting:		
Basic procedure	3-4	3-2
Scope	3-3	3-2
Symptom index	3-5	3-3
Table		3-4
Truck characteristics:		
Brake system	1-15	1-17
Fuel system	1-11	1-13
Steering system	1-16	1-18
Truck loading:		
Gross axle weight rating (GAWR)	1-18	1-19
Gross vehicle weight rating (GVWR)	1-18	1-19
Truck maneuvers:		
Forward	2-9	2-54
Reverse	2-11	2-54
Stopping	2-10	2-54
Truck operation, unusual conditions:		
Cold weather	2-42	2-88
Fording	2-45	2-91
Hot weather	2-43	2-89
Snowy or icy terrain	2-42	2-89
Transfer ranges	1-13, 2-44	1-15, 2-90
Unusual terrain	2-44	2-90
Turn signal indicator	2-17	2-59

u

Underwood heaters, winterization kits, operation	2-36(a)	2-80
Unusual terrain:		
Maintenance after operating on	3-10	3-14
Operation	2-44	2-90
Use, preparation for	2-3	2-3
Utility, truck	1-6(e)	1-4

Index 11

TM 9-2320-289-10

INDEX

SUBJECT	Paragraph	Page

V

Vents:
    Air exhaust 1-7   1-7
    Air, operation 2-22   2-62
Voltmeter ... . 2-2(12)   2-2

W

    Warning flasher, hazard, operation 2-21   2-62
Warranty 1-4   1-1
Water, fording:
    Fresh, maintenance 3-11   3-15
    Salt, maintenance 3-11   3-15
Weapons mount operation.. 2-24   2-64
Weather:
    Extreme cold, maintenance 3-8   3-14
    Extreme hot or humid, maintenance 3-9   3-14
Weight classification marker 1-7   1-6
Weight limits 1-18   1-19
Wheels and tires, changing tires 3-6   3-9
Window, tailgate:
    Location 1-7   1-8
    Operation 2-23   2-63
Window crank, tailgate:
    Location 1-7   1-8
    Operation 2-23   2-63
Windshieldwiper/washercontrol, operation 2-18   2-60
Winterization kit:
Defroster\interior heating ....2-36(b)   2-81
    Differences between models....................2-36(d)   2-82
    Enclosed cargo area personnel heater, operation .... 2-36(c)   2-82
    General description   1-20
                                                                   1-19   2-80
    Underwood heater, operation ....2-36(a)   1-17
Wiring and lighting system ....1-14(a)

By Order of the Secretaries of the Army, the Navy and the Air Force:

JOHN A. WICKHAM, JR.
General, United States Army
Chief of Staff

Official:

R.L. DILWORTH
Brigadier General United States Army
The Adjutant General

CHARLES A. GABRIEL, *General, USAF*
Chief of Staff

Official:

J. J. Went
Lieutenant General, USMC
Deputy Chief of Staff for Installations and Logistics

Distribution:
To be distributed in accordance with DA Form 12-38, Operator maintenance requirements for Truck, Commercial Utility Vehicle Cargo, Tactical 4x4, M1008, M1008A1, M1009, M1010, M1028, M1031.

☆U. S. G. P.O. 1986-644-003

LINEAR MEASURE

1 Centimeter = 10 Millimeters = 0.01 Meters = 0.3937 Inches
1 Meter = 100 Centimeters = 1000 Millimeters = 39.37 Inches
1 Kilometer = 1000 Meters = 0.621 Miles

WEIGHTS

1 Gram = 0.001 Kilograms = 1000 Milligrams = 0.035 Ounces
1 Kilogram = 1000 Grams = 2.2 Lb.
1 Metric Ton = 1000 Kilograms = 1 Megagram = 1.1 Short Tons

LIQUID MEASURE

1 Milliliter = 0.001 Liters = 0.0338 Fluid Ounces
1 Liter = 1000 Milliliters = 33.82 Fluid Ounces

SQUARE MEASURE

1 Sq. Centimeter = 100 Sq. Millimeters = 0.155 Sq. Inches
1 Sq. Meter = 10,000 Sq. Centimeters = 10.76 Sq. Feet
1 Sq. Kilometer = 1,000,000 Sq. Meters = 0.386 Sq. Miles

CUBIC MEASURE

1 Cu. Centimeter = 1000 Cu. Millimeters = 0.06 Cu. Inches
1 Cu. Meter = 1,000,000 Cu. Centimeters = 35.31 Cu. Feet

TEMPERATURE

5/9(°F − 32) = °C
212° Fahrenheit is equivalent to 100° Celsius
90° Fahrenheit is equivalent to 32.2° Celsius
32° Fahrenheit is equivalent to 0° Celsius
9/5(°C + 32) = °F

APPROXIMATE CONVERSION FACTORS

TO CHANGE	TO	MULTIPLY BY
Inches	Centimeters	2.540
Feet	Meters	0.305
Yards	Meters	0.914
Miles	Kilometers	1.609
Square Inches	Square Centimeters	6.451
Square Feet	Square Meters	0.093
Square Yards	Square Meters	0.836
Square Miles	Square Kilometers	2.590
Acres	Square Hectometers	0.405
Cubic Feet	Cubic Meters	0.028
Cubic Yards	Cubic Meters	0.765
Fluid Ounces	Milliliters	29.573
Pints	Liters	0.473
Quarts	Liters	0.946
Gallons	Liters	3.785
Ounces	Grams	28.349
Pounds	Kilograms	0.454
Short Tons	Metric Tons	0.907
Pound-Feet	Newton-Meters	1.356
Pounds per Square Inch	Kilopascals	6.895
Miles per Gallon	Kilometers per Liter	0.425
Miles per Hour	Kilometers per Hour	1.609

TO CHANGE	TO	MULTIPLY BY
Centimeters	Inches	0.394
Meters	Feet	3.280
Meters	Yards	1.094
Kilometers	Miles	0.621
Square Centimeters	Square Inches	0.155
Square Meters	Square Feet	10.764
Square Meters	Square Yards	1.196
Square Kilometers	Square Miles	0.386
Square Hectometers	Acres	2.471
Cubic Meters	Cubic Feet	35.315
Cubic Meters	Cubic Yards	1.308
Milliliters	Fluid Ounces	0.034
Liters	Pints	2.113
Liters	Quarts	1.057
Liters	Gallons	0.264
Grams	Ounces	0.035
Kilograms	Pounds	2.205
Metric Tons	Short Tons	1.102
Newton-Meters	Pound-Feet	0.738
Kilopascals	Pounds per Square Inch	0.145
Kilometers per Liter	Miles per Gallon	2.354
Kilometers per Hour	Miles per Hour	0.621

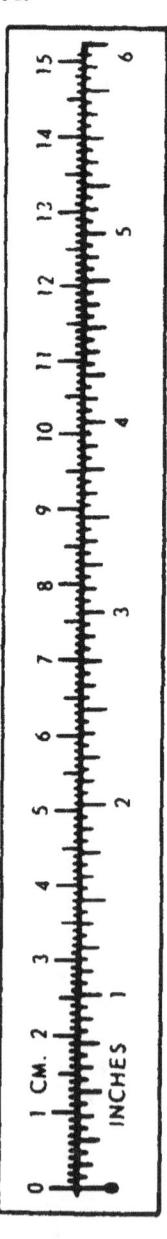